To Soar with the Tigers

The Life and Diary of Flying Tiger, Robert Brouk
2nd Edition

Jennifer Holik with Robert Brouk

Copyright Information

Copyright © 2013 Jennifer Holik.

Publisher: Generations, Woodridge, Illinois

Editor: Marjorie Sucansky
Cover Designer: Jennifer Holik

Holik, Jennifer, 1973 –
　　　To Soar with the Tigers : The Life of Flying Tiger, Robert R. Brouk / Jennifer Holik.
　　　　　Includes bibliographical references and index.
　　　　　ISBN 978-1-938226-26-7

This book is dedicated with love to
Virginia S. Davis and the late Robert R. Brouk.

Purpose

The goal of this book is to illuminate the brief, shining life of Robert Brouk. It includes his role as a pilot in the American Volunteer Group in the months prior to Pearl Harbor, until disbandment in July 1942, and his role as an instructor in the beginning of World War II.

Acknowledgments

This book could not have been written without the support and assistance of many people beginning with Robert's widow, Virginia S. Davis. Without her support, love, encouragement, stories, and photographs, Robert's complete story would not have been told.

There are many other people I would like to thank: My editor, Marjorie Sucansky; Patricia Holik for helping me choose a title; the late Richard Holik for "introducing" me to Robert years ago; the late Charles R. Bond, Jr., and the late John Richard Rossi for corresponding with me in 1997 about their memories of Robert; Charles Baisden, 3rd Squadron Armorer in the AVG, for sharing his memories of Robert and answering many questions during the course of my research; Jo Neal, President of AVG-FTA for answering numerous questions for me; Brad Smith for providing a photograph taken by his father, R.T. Smith, of Robert in 1942; Karen Halla and Mary Ellen Jenicek with the Morton East High School Archives, who provided information and several photographs of Robert from high school; John Gieger at the Hawthorne Works Museum, located at Morton Junior College in Cicero, who helped me locate information not only from the Hawthorne Works archives but also from the Morton Junior College archives.

Note: Robert's War Diary appears in this book verbatim from his original diary, but spelling and punctuation have been revised where necessary for clarity. Editorial comments have been placed within the diary to expand on events Robert described in his diary.

Table of Contents

Biography
1917 – 1941

Robert Ralph Brouk was born September 2, 1917, in Oak Park, Illinois. His father, Peter Brouk was a Bohemian immigrant, arriving in the United States January 5, 1900, at the age of 10.[1] His mother, Emily Novak, was born in Chicago to Bohemian immigrants. Peter and Emily had three children, Peter, born September 4, 1915, died April 1, 1922[2]; Robert; and Harold, born October 6, 1923, died December 23, 1983[3]. Harold joined the United States Army during World War II and was sent to Fort Warren, Wyoming, for training.[4] During World War II, Fort Warren served as a training base for U.S. Army Quartermaster Corps Soldiers.

Robert's father Peter was a sign painter,[5] and took great pride in his community. As a business owner, he was a member of the 22nd Street Business Men's Association in Cicero, where he served in the role of Secretary in 1931. The organization's goal was to revitalize and encourage growth of the businesses along 22nd Street in the villages of Cicero and Berwyn in the 1930's.[6]

As a child, Robert attended Woodrow Wilson grammar school. From 1931 until 1935, he attended J. Sterling Morton High School in Cicero, where he was active in campus clubs and sports. Robert participated in the Drum and Bugle Corps his freshman,

Yearbook photo - J. Sterling Morton Drum and Bugle Corps 1932
(Courtesy Morton East High School Archives)

Yearbook photo - J. Sterling Morton 1934 Wrestling Team
Robert is on the top row, fifth from the right
(Courtesy Morton East High School Archives)

Yearbook photo - J. Sterling Morton 1935 Wrestling Team
Robert is on the bottom row, fourth from the left.
(Courtesy Morton East High School Archives)

He also served as Vice-President of the Chemistry club his senior year; and played Intramural Sports his junior year.[7]

J. Sterling Morton 1935 Chemistry Club
Robert is on the top row, first person on the right
(Courtesy Morton East High School Archives)

Robert graduated from Morton High School in Cicero, Illinois, in 1935.[8] After graduation, he attended Morton Junior College, in Cicero, from 1935 – 1937. While attending Morton during his freshman year, Robert participated in the Chemistry Club, Engineer's Club, and continued with his wrestling.

The Engineer's Club was a very active club which took field trips to engineering departments at the University of Illinois and Purdue, the Argo Starch Company, and Universal Oil company refinery, and other similar companies. The Morton Junior College club was part of a larger group, the Midwest Engineers' Club which consisted of five other area junior colleges' clubs. This allowed ideas to be shared freely for everyone's benefit.

The Chemistry Club also took field trips to chemical laboratories

Seated: E. Seyk, A. Jantar, S. Bourbaki, G. Thomas, C. Hosek, C. Wankat, E. Stantejsky, C. Sellen, S. Solopolous, V. Lameroux, C. Benes, J. Vendel. First Row Standing: E. Mosna, M. Molik, H. Grillot, J. Koveckis, T. Erdelyan, R. Brouk, H. Kerber, E. Sisul, J. Javorsky, O. Vasak, H. Vyletal, J. Vltek, R. Burg, J. Braun, M. McIntyre, A. Buboly, H. Gass. Second Row Standing: A. Novy, J. Koneeny, A. Wild, G. Tugana, E. Wassel, M. Kokoska, K. Brensten, M. Hlinsky, S. Palansky, W. Smaus, E. Kanak, J. Martinek, A. Yuska, O. Warning, H. Widiger, W. Plourd, E. Michl.

Yearbook photo – Morton Jr. College 1936 Engineers Club
(Courtesy Morton Jr. College Archives)

of area colleges and Abbott Laboratories, in addition to conducting experiments and sharing ideas. Robert served as the secretary-treasurer during his freshman year.[9]

Fit in mind and fit in body, Robert was mentioned in a Morton wrestling article, in the *Morton Collegian*, because the college had the largest squad in years, and "Competition rife in the 135 lb. division with Robert Brouk…"[10] seems to speak to Robert's competitiveness and skill. The Morton Junior College Wrestling Team exceeded expectations in 1936 by winning against several four year colleges in the Chicago area.

Also during 1936, Robert was inducted on 21 June, as a Master

Yearbook photo – Morton Jr. College 1936 Wrestling Team. Robert is on the bottom row, middle. (Courtesy Morton Jr. College Archives)

Builder in the Cicero Chapter, No. 12, Order of the Builders.[11] The Order of Builders is still today, a part of the Masons, for boys ages 9 to 21, and its purpose is to teach the members the principles of democracy and ideals of Freemasonry, while participating in various social, athletic and civic projects. To be inducted, the boy must be a close relation to a current Masonic member. Robert's father, Peter Brouk, was a Mason in Cicero.

During Robert's sophomore year at Morton College, he again participated in the Chemistry Club and the wrestling team.

The Wresting Team, in its second successive season, did not win as many matches as it had the year before. However, Robert was listed in the Pioneer Yearbook as one individual who consistently won the matches in which he competed.[12]

In the *Morton Collegian* Prophecy article about the sophomore graduating class, it was predicted that Robert would be, "Research expert for the We Chew Your Gum Co,"[13] which speaks to his probable sense of humor. In the Pioneer Yearbook for 1937, it lists Robert Brouk as a Pre-Engineering Student who was, "Athletic, scholastic and sociable."[14] Robert graduated from Morton Junior College on 13 June 1937.

Yearbook photo – Morton Jr. College 1937 Chemistry Club.
Robert is on the bottom row, middle.
(Courtesy Morton Jr. College Archives)

Yearbook photo – Morton Jr. College 1937 Wrestling Team.
Robert is on the bottom row, middle.
(Courtesy Morton Jr. College Archives)

After graduating from Morton Junior College in 1935, Robert attended Lewis Institute of Technology. It was while attending Lewis in 1939, that Robert joined the Army Air Corps.[15] His parents, Peter and Emily moved from their home in Chicago, to 2120 59th Court, Cicero, after 1930. This is the home to which Robert returned after his service with the American Volunteer Group.

Robert graduated from the United States Army Air Corps Advanced Flying School on 30 August 1940, at Kelly Field, Texas. Upon graduation, Robert received his commission as a Second Lieutenant in the Army Air Corps Reserve.[16]

Later, while serving at Mitchel Field, Robert learned about the formation of the American Volunteer Group (AVG.) Mr. 'Skip' Adair of Central Aircraft Manufacturing Company (CAMCO) visited Mitchel Field to recruit pilots, mechanics, and armorers to join the AVG in China. The AVG was being formed by all volunteer pilots and ground crew under the leadership of Claire Lee Chennault.

In 1937, Chennault was a retired Captain of the U.S. Army Air Corps. He was working under the leadership of Madame Chaing Kai-shek between 1937 and 1941 to help China build an air force similar to the U.S. structured Army Air Corps. China was on the verge of war with Japan and needed assistance building a solid program and defense system.

After consulting for many months with the Chinese Air Force, Chennault handled the strategic planning of air raid systems, air fields, and pilot training. He flew many missions with the Chinese pilots against the Japanese and developed new combat strategies. He took this knowledge and proposed the formation of the AVG.

In Chennault's mind, the purpose of the AVG was to defend China and Burma's main roads and supply routes from Japanese attack. He also wanted to attack Japanese staging areas and supply depots. To accomplish this, he needed a well-trained group of pilots and ground crew.

Without support from the U.S. Army Air Corps or U.S. Navy, who did not want to see volunteer American pilots in China, Chennault enlisted the aid of President Roosevelt. The president sent out an unpublished executive order on April 15, 1941, encouraging pilots and ground crew to join the fight. These men would be honorably discharged from the U.S. Army Air Corps or Navy, to join the AVG and were hired by the Central Aircraft Manufacturing Company (CAMCO.) The president also authorized specific individuals to visit air fields to recruit pilots and ground crew. Mr. 'Skip' Adair of CAMCO visited Mitchel field in late April and early May 1941 where Robert was stationed. It was at here that Robert decided to join the AVG.[17]

Why were the men of the AVG known as the Flying Tigers? Chennault explained it was simply because they painted the shark toothed design on the noses of their planes. This design was not original to them, but something they copied from a R.A.F. squadron that served in the Libyan Desert. Walt Disney's company then created insignia of the plane flying through a 'V' for Victory. The design and name caught on and people all across the country were calling these men the Flying Tigers.

This story is the history of the Flying Tigers from the words and perspective of Robert Brouk. Robert proved to be an avid recorder of events and kept a diary of his experiences during his time as a pilot in the American Volunteer Group. His diary provides a graphic account of events in China at the beginning of the war. His diary is presented unchanged and in its entirety with editors comments bracketed.

Biography Notes

[1] "New York Passenger Lists, Roll T715_99: 1820-1957." Database *Ancestry.com*. (http://www.ancestry.com : accessed 5 May 2010), entry for Peter Brouk, age 11, arrived New York, New York, 1900 [Noordland].

[2] Florida Office of Vital Statistics, death certificate 23196 (1942), Robert R. Brouk; Bureau of Vital Records, Tallahassee.

[3] "Social Security Death Index." Database *Ancestry*.com (http://www.ancestry.com : accessed 25 September 2010), entry for Harold Brouk, 1983, SS no. 336-18-0271.

[4] City of Chicago, Illinois, probate case files, no. 43P 1805, Robert R. Brouk (1943), Petition for Letters of Administration, 12 March 1943; Circuit Clerk's Office, Chicago.

[5] "1910 United States Federal Census Roll T624_254 : 1910." Database *Ancestry.com*. (http://www.ancestry.com : accessed 3 September 2010), entry for Peter Brouk, Chicago, Illinois.

[6] "Richard Kauffman Heads Cicero Business Group," *Chicago Daily Tribune (Chicago)*, 8 March 1931, p. H5; digital images, ProQuest (http://www.proquest.umi.com : accessed 3 June 2010), Historical Newspaper Collection.

[7] "J. Sterling Morton Year Book 1935." Database *Ancestry.com*. (http://www.ancestry.com : accessed 1 May 2010), entry for Robert Brouk, Cicero, Illinois.

[8] Karen Halla, Cicero, Illinois. [(E-address for private use),] to Jennifer Holik, email, 10 May 2010, "Re: Historical Society of Cicero," Robert Brouk Correspondence File, Robert Brouk Book Research Files; privately held by Jennifer Holik [(E-address) & street address for private use,] Woodridge, Illinois.

[9] Chemistry Club column and photograph, in Pioneer Yearbook, ca. 1936, p. 66; Held by Morton Junior College Library, [3801 S. Central Avenue,] Cicero, Illinois, 1936.

[10] "Wrestlers Have Largest Squad For Many Years," *(Cicero) Morton Collegian,* 6 November 1936, p. 4, col. 4.

[11] "Cicero Chapter of Builders to Induct Leaders," *Chicago Daily Tribune (Chicago)*, 21 June 1936, p. W7; digital images, ProQuest (http://www.proquest.umi.com : accessed 3 June 2010), Historical Newspaper Collection.

[12] Wrestling Club, column and photograph, in Pioneer Yearbook. ca. 1937, p. 66;. Held by Morton Junior College Library, [3801 S. Central Avenue,] Cicero, Illinois, 1937.

[13] "Class Prophecy," *(Cicero) Morton Collegian*, 28 May 1937, p. 2, col. 4.

[14] Robert Brouk sophomore photograph, in Pioneer Yearbook, ca. 1937, pg 91; Held by Morton Junior College Library, [3801 S. Central Avenue,] Cicero, Illinois, 1937.

[15] "Untitled," *(Berwyn) Berwyn Life*, 11 November 1942, p. 1, cols. 4-5.

[16] "Nine Young Chicagoans Receive Commissions as Lieutenants in Army Air Corps Reserve :Fifth Recent Class. " *Chicago Daily Tribune (Chicago)*, 31 August 1940, p. 6; digital images, ProQuest (http://www.proquest.umi.com : accessed 3 June 2010), Historical Newspaper Collection.

[17] "Tiger Diary," *(Chicago) Herald-American*, 26 July 1942, p. 1.

"What's Next" War Diary of Robert R. Brouk April 1941 – July 4, 1942

April through June 1941

During April and May of 1941, a Mr. 'Skip' Adair of CAMCO, came to Mitchel Field and talked to the fellows in regards to flying in China for CAMCO, mainly to protect the Burma Road. After much hullaballoo and studying pros and cons, and finally getting all the Army matters straightened out, a group of 11 of us left Mitchel around the first of June. There were Atkinson, Dupouy, Harris, Sawyer, Walroth, Little and myself from the 33rd Squadron. Kelleher, Olson, Cook and Martin were the others. We were to report to Los Angeles the 12th of June. The first group of mechanics from Mitchel sailed about the first week of June.

I arrived home in Cicero on Memorial Day and due to a hold-up in sailing, was notified to report the 18th. My mother, brother, cousin and I left Cicero on the 9th of June by car. We drove through the Southwest, stopping at main points of interest, such as Will Roger's Memorial, Painted Desert, Grand Canyon, Petrified Forest and arrived in Los Angeles the 14th. After reporting on the 16th at Harlow Aircraft in Alhambra, I learned sailing was again delayed. Those fellows already present were staying at the exclusive Johnathan Club in L.A. so I proceeded to do likewise. My family left for home on the 20th.

July 1941

After an interesting stay in L.A., seeing points of interest and going to places of amusement, along with 30 other early arrivers, we left L.A. at 6:30 A.M. Sunday, July 6th for San Francisco, which was traveled to in about 12 hours. There we met the rest of the group. On the 7th, about half of us were given quarters on our ship, the *Jagersfontein,* and the rest came aboard on the following day. The group consisted of approximately 50 pilots, both Army and Navy, and 73 Mechanics and other enlisted men.

Sailing from S.F. was at 1:20 P.M. Thursday, July 10th. The first day out, we ran into a small swell which enabled us to get our sea legs and ascertain *(sic)* our stomachs to the toss and roll of the sea.

From the first night, we practiced blacking out and were very strict to its adherence.

The second day out was rather calm so our most delicious food was thoroughly enjoyed. A boat drill was also held in the morning.

On Sat., the 12th, Dr. Samuel Pan, a fellow traveler along with 17 other Chinese exchange students returning from the U.S., started a Chinese language class which a group of us started.

Sunday started a group of lectures by the Chinese travelers on China and customs of the Chinese, which they thought would prove interesting and helpful to us. That evening an informal quiz program was held in the Social Hall. Beer served for prizes.

Monday's warmer weather brought into use the outdoor pool which many of the fellows made quick use of.

Hawaii – the land of hula and palm trees. It came into view Tuesday morning and the port was entered about 9:30 P.M. The passengers had deserted the ship by 10:30. All were in to see or do *(sic)* the City of Honolulu. The next morning was also spent in part, and not having much time, I did well to see Waikiki Beach and do a little necessary shopping. We sailed at 1:00 P.M.

The next morning found us out of sight but not memory of Hawaii, but in sight of two Navy cruisers; which were to act as convoy for the journey. The lectures and classes in Chinese continued as the next stage of the journey began in earnest.

Friday was needle day. We all received our first shot for cholera and also smallpox or vaccination. Our course was almost due South which caused a lot of prophesying on our next stop, but nobody was "in the know."

The next week passes as prototype as the usual day. *(sic)* Breakfast around eight, then a thorough sun bath on the deck, sometimes reading and interrupted around 11:00 for the morning cool, refreshing drink served by the barefooted turbaned Javanese boys. A little side light is that these boys, around 33 years of age, make only one of *(sic)* two trips a year as they can live on $30.00 for almost six months. Their pay is from tips from the passengers. Most of them cannot speak any English, but are well trained in their jobs. Luncheon is at 1:00 for the officers, enlisted men ate at 12:00 noon. The luncheon consisted of a vegetable salad, soup, entrée, dessert and fruit. The coffee was served afterward in the Social Hall.

From two till four was Chinese language class. I'm taking it with the hope of using it, but like any other language, the further you progress, the more intricate and complicated it becomes. Class was

Bob on the Jagersfontein en route to China.
(Courtesy Virginia S. Davis)

usually ended by the "boys" again making the round of the ship serving tea and cookies.

From four till seven was usually spent in reading, especially books associated with China, Japan, or its associated complexities.

The most interesting episode during these last few days was the conversion of "pollywogs" to "shellbacks". This is a time honored custom of the many of initiating all new sailors who cross the equator for the first time. It consists of a group of Shellbacks, those who have already crossed, acting as prosecutors and administrators of justice. This administration of justice consisted of a thorough shellacking of the "body stern," coloring of the body, a hit of pungent fish, shave with a wooden razor, and finally a "helped" entrance into the pool with a triple ducking for good credit. It was fun watching and participating even though it left its memorial marks.

The evening meal was not to be outdone. With olives and celery followed by salad, fish dish, the entrée, dessert, fruit and again coffee in the Social Hall.

Most evenings were just spent sitting on the darkened deck with only the blue lights showing the lifeboat stations, and the occasional glow of a cigarette. No lights were permitted to be shown, as were matches forbidden to be struck outside. Many are the fellow who blindly ran into a deck chair or a fellow passenger. But safety is safety.

August 1941

Somewhere above Australia we slept through a day. That is, went to bed on Friday and woke up on Sunday. That and the fact we changed two U.S. Cruisers for one Dutch cruiser on the 3rd of August was the only bit of excitement for the past week.

For the first time since leaving, we had a "no blackout" night. With our escort we anchored off a small island and picked up an Australian Straits Pilot. The next day we passed many small islands in the afternoon passed close by a fortified entrance to the opposite end of the Straits of Torres. There we left off our Pilot and continued on our journey.

Since then we have been seeing scattered islands on both sides of us, each fellow expressing his guess to its name and position.

My daily routine now usually consists of two hours Chinese class in the afternoon; the sun bathing in the morning with about an hour of elementary hand balancing or wrestling just before noon. The latter part of the afternoon is spent reading.

This time also marked the end of picture taking for a while as all cameras had to be turned into the purser's office. It seems all the governments don't relish photos of their islands' fortifications turning up in unwelcomed hands, so – no pictures!

Journey still moving along at its 15 knots, more islands, more guesses as to their position and names. One, somewhere very near Java or Bali, we spent the morning about three miles off shore while the Dutch Cruiser "Jana" went in closer and oiled up. Our first indication of our first stop was posted on Saturday the 11th, where it stated we would dock early Monday morning in Singapore and should have our baggage ready for disembarkation.

Monday, August 11, we arrived outside the harbor of Singapore about 7:00 where the quarantine officer and pilot came aboard. We finally docked around 8:00 but due to a little difficulty with the immigration department, we did not leave ship until after 2:00 P.M.

After spending almost an hour trying to declare money, mail, get a return pass (which we finally found out was not needed) we left the gate into the Humdrum of Singapore; the melting pot of the Orient. Our first bit of orientation was a short ride on the "Tram," then came the long waited surprised(sic); a ride in a real rickshaw pulled by a typical Chinese cootie.(sic)

That ride was about the most interesting and enlightening ride I have ever taken. It started out in a typical Chinese section with the stores running into each other; their wares almost intermingled and their odors certainly were. Most of the shop keepers were young Chinese and in a few cases, the shops were run by other Orientals than Chinese; somewhat sconitic(sic) in appearance. The trail then ran along these streets, strewn literally with humanity; Indians with their skirts and turbans, Chinese women with their babies on their backs in papoose style, the Sheiks or Indian policeman who are almost all over 6 feet and real strapping fellows. Here and there were whites doing either business or rubber-necking; every ten feet it seems the coolies had to dodge a stunted automobile. Here most cars are of European; mainly British built, and are small; comparable to the Austin. All cars here have right hand drive and travel along the left hand side of the road. I never did get used to looking in the opposite direction before crossing the street,

*Robert sightseeing in China
(Courtesy Virginia S. Davis)*

although it seems as if no one else does from the manner in which jay walking is carried on.

Soon we came into the more modern section with its newer buildings, larger, cleaner and more spectacular window displays. Here also the traffic became much more dense and dodging really came into being. A river, or rather Singapore Canal, was crossed and traffic on this canal was literally stopped by the great number of junks and sampans reaching from one bank to another.

Another prevalent means of transportation in Singapore is the bicycle. There are almost no women riding them, but every male from young boy to the old Indian with his skirt and turban fluttering in the breeze is ducking in between rickshaws and pedestrians. First comes the automobiles, then the rickshaw, next the bicycle and finally the poor pedestrian. He hardly stands a chance.

Besides all the shops and interesting shop keepers, another human took up quite a bit of notice.

The soldiers in Singapore almost seem to outnumber the civilians. All kinds of nationalities seem to be represented. The Indians, Australians, New Zealanders and even quite a few Brishers*(sic)* are noticed strolling along. There are also quite a number of sailors, although now they were mostly at sea for any emergency.

Our little ride ended in front of a curio shop where we entered and did a little rubber-necking ourselves. There were very many intricate and delicate pieces of hand carved wood made out of a single piece. Also many pottery, *(sic)* china and silverware. This place did not happen to have any jade though.

A short walk back to the post office to purchase some stamps and then we found a cab driver; a Malayan who spoke English. We hired him and his 10 H.P. Ford, to drive us around town and show us some of the interesting sights. We passed the famous Raffles Hotel; but more of that later. Saw several convents and girls schools; thru more real Chinese sections and out into the suburbs where we came across new, modern, larger and beautiful homes. These belonged mostly to Consular workers or rich businessmen. We also saw a few of the Army Officers' homes, which were very nice. Finally we came to the Botanical Park. It is a large, very well kept park containing many variety of plants, trees, shrubs and flowers. It even had its water lily pond and to my surprise, a large number of monkeys. The monkeys were rather tame and would take peanuts out of the hands of their feeder. Every one bought peanuts and fed the little fellows. We did happen to see a young mother monkey with her offspring clinging to her chest as she scampered from spot to spot or partook of peanuts we fed her.

Our next stop of interest was at the Tiger Balm Museum. It is a fully furnished house, unlived in of course, but richly decorated with real Chinese decoration. All the rugs, tapestry, and handiwork was*(sic)* very colorful and pretty. Throughout the house were innumerable Jade carvings. That alone must have amounted to over a million dollars. From there we drove out to the Civil Airport; now taken over by the R.A.F. We talked to a young New Zealander, R.A.F. Sergeant. We found that a friend of Howard, who was my traveling companion, was stationed here as an observer, but was not there at the present. So we stopped for a cocktail before we left for the Raffles Hotel for dinner. Upon walking in, the first person we saw was Howard's friend, Price, and several of our bunch so we promptly joined them. From there it was back to the Airport Restaurant for dinner and then out for the evening.

We went to the famous New World, which is somewhat of a Chinese amusement park. It contains a cabaret, several movie houses, and even more outdoor play houses where we saw real Chinese drama being presented. One interesting sidelight is that there are not acts. It is continuous and when they want to change scenery, the actors turn their back to the audience and young boys change the scenery. Also only men take part in these plays and young boys take the high pitched women's parts. We looked around a bit and then went to the cabaret where we had a few dances with the Chinese hostesses. These girls are mostly all pretty, speak English and highly prize these jobs. From there we went back to Raffles for the last few minutes before the 12:00 curfew on bars. There we met more of our group along with three R.A.F. pilots who saw service over London; all had several confirmed victories. And so then back to the ship via an open air midget cab.

The next morning we loafed close to the boat, buying bananas, milk, ice-cream, cigarettes and sun hats from the traveling Chinese Merchants.

We left the dock at 2:15 P.M. with two of the group missing. As we were clearing the pier, they came up and quickly went to the end of the long pier and got a taxi boat. They came aboard outside the harbor as the harbor pilot left.

And so on our way to Rangoon.

The mouth of the Irrawaddy River came into sight on the afternoon of the 15th; we picked up our pilot and started our trip down the river to Rangoon. As soon as we entered the river, it started raining ala Burma. It soon became dark so*(sic)* with the rain and darkness we had our last night's sleep on the *Jagersfontein*.

At 5:30 A.M. the breakfast gong awoke us for the last time. After a short breakfast we transferred to a lighter and were taxied to the wharf among the numerous water taxies and sampans.*(sic)*

A short bus ride and we were dropped at the Silver Brill for a fine breakfast. From there we had a quick ride through the quaint streets of Rangoon, somewhat like Singapore, to the railroad station.

Amongst the many Indians and Burmese, we sorted our baggage and broke up into groups of six for a second class railroad compartment. On our narrow gage, small engine train, we started a very interesting ride at 1 P.M. Saturday, August 16th.

The coach itself was rather small with two fixed seats on each side with the center section folding out from the side of the coach to form a bed or place for three during the day. A drop bed which laid flat against the side and made a double bed above the seat bed, so the compartment held six day or four night passengers. They were not very comfortable, but sufficed for the short journey.

The train made very good time and rode very smoothly. The railway ran through many villages, some rather large with railway sidings, a large station; somewhat of a main street and a few larger stores. Other villages were nothing more than a few bamboo homes and a mud road. All along were numerous rice fields, all under water and many natives pulling out the ripe stalks, none *(sic)* stopping to watch the train; others going on nonchalantly with their work.

The peoples passed were mostly Burmese *(sic)* or Indians. They all wear their dress like cloths wrapped around their lower portions and a shirt of various colors and designs. The Indians wear turbans and the Burmese and a few Chinese have larger bamboo sun hats. The land was very flat and as afar as the eye could see were rice fields and a few spots of thick jungle-like growth of shrubs and trees. Many large and small birds were seen throughout the trip, large swan-like birds called "Paddy Birds" were seen on all the rice fields. A few larger, black buzzard birds were also seen. Water buffaloes were as prevalent as cows back home. They would lie in the water ditches along the rice fields, irrigation canals, rivers, and small puddles of water.

Another object of interest all along the way were the numerous idols. Some were large busts of Buddha, as high as 50 feet, enclosed by brick walls with only the face showing, and the front section open to show the whole front of the bust. Other objects were intricate towers topped by gold kraft *(sic)* work. Some were a part of a large bird, who sat on the ground like a boat. These birds were about 100 feet long and the towers about 30-50 feet high.

We stopped at about 3 o'clock at a larger village and were each given a lunch box, which was not too appetizing, but proved rather welcomed. At this same stop, some of us took off into the village proper. We saw a native having a shave with a straight razor and no lather whatsoever, just dry, and as clear a shave as you would want. Whenever we came into view all the women would run away or hide their faces. The other populace would just stare back. Most of the houses were upon piles above the water which was everywhere, but on the road. The animals lived along with their owners in the same rooms.

The Burma Road, which we followed all the way from Rangoon to our destination also ran through this town, so we saw several convoys of American trucks pass thru on their way probably to China.

Our next stop several hours later brought us another group of station well-wishers, who seem to spend all day at the stations, either just watching, selling their wares, or begging. The small children begged for money and would scramble when coins were tossed to them. We saw a few Buddha priests in their yellow robes and shaven heads.

At about six o'clock we came to Toungoo where we saw several fellows who were already at the outpost. From there it was only a few minutes ride to where we stopped at a small village and our baggage unloaded into trucks and we were driven in station wagons and Plymouths ('41) about eight miles to the R.A.F. Field.

So after 36 days on a boat, a bus ride, train ride and a station wagon ride, a trip half way around the world was at a halt. From sunny California to rainy Burma. Blackout ship, rickshaw rides, strange people and customs; these all filled the otherwise dreary days.

The field here, built by the R.A.F. in the spring of '41, is well laid out for an air post. It was cleared from semi-jungle into a large scattered post. The roads are of stone and clay, hand made by the coolie workers. The buildings are of frame construction with laced bamboo matting for sides and roof. The builders were at least 1000 yards apart. The officers' quarters were of the barracks type; one long room with 16 mosquito net covered beds, a dresser between each set of two; there was a shower room 50 feet away as well as separate springs again 50 feet. The water was chlorinated at the well before distribution so it was all drinkable. Of course, there was electricity, but no radio as yet. A phonograph served for entertainment, in our particular barracks. The other officers' barracks as well as enlisted men's were the same.

There were separate mess halls for officers and enlisted men. The hangar was almost two miles from the mess hall; so it was either a long walk or a bumpy ride in one of the station wagons, of which there were about three and also three '41 Plymouths.

The hangar was of timber and sheet metal; capable of about 18 planes. *(sic)* Scattered around the hangar were bomb proof shelters as well as dugouts for planes. The petrol dump was about a mile from the hangar; between the mess hall and hangar.

There was a part exchange, laundry, barber and other small conveniences on the port, so we were not altogether away from civilization.

Since we arrived during the rainy season, we were well initiated. Black clouds would soon appear, a heavy rainfall, a sudden stop; this would continue for hours.

The meals became more inviting, or we became more accustomed to the food served in Burma. The days were usually well spent with a few details to be taken care of at the hangar, an evening spent in Toungoo, exercise or sunning during the day.

Robert in Burma.
(Courtesy Virginia S. Davis)

I was supposed to leave for Rangoon on Friday, August 22, but no planes were ready for ferrying, so my packing went for naught. The same Saturday, *(sic)* but Sunday we did leave here about 11:30 in the Beechcraft with a Major of the Chinese Air Force as pilot. The weather was rather poor with mostly instrument flying.

Since it was Sunday, after we arrived, there was no one waiting for us, so we had a quarter mile walk to where we caught a local Rangoon bus, filled with its usual melee of passengers; Indian,

Burmese, and others unbeknown to me. About half way into town a '41 Chevy pulled up behind the bus and honked the horn. It was the company car which missed us at the airport and caught up to us. From there, we were taken to Minton Mansions. It was supposed to be the second best hotel in town; but was rather old and not too pleasantly decorated or furnished; but offered a suitable place for sleeping quarters. After a quick shower and change of clothes, a not too delicious or appetizing luncheon, we drove into town to the Strand Hotel where Mr. Pawley was living. We discussed a few points of discussion; had a drink with him, then one at the bar and proceeded to see some of the town. We had use of Mr. Pawley's car for the day, so we drove, or had our Burmese driver drive us to the Shwe Dagon Pagoda just outside of town.

It is one of the largest Pagodas in the world, covering about four square blocks, surrounding a 100 foot, gold plaited shrine. This pagoda is almost 2500 years old, built over hairs of Buddha's head. There are numerous smaller shrines which are gifts of private families, and depending upon their wealth is the many jewels or richness of shrine. *(sic)*

All visitors must go through this Pagoda barefooted; so after we got through, we stopped at a pharmacist on the way back and bought Potassium Permanganate to bathe our feet in.

After this very interesting sight into the religious nature of Burma and Buddhism, we proceeded back to the hotel for dinner. We started out for the theatre, but found out the show did not start till 9:30; so back to the Strand for a drink and then to bed.

The next day was spent in town shopping and finding out some information regarding passage back to the states for some of the fellows who contemplated going back home.

It was raining too hard to fly back so the afternoon was spent with a Mr. Gibson who was a former Navy pilot, but is now a pilot for Brewster Co., and testing the R.A.F. "Buffalo" here at Rangoon. That evening we did make the show and saw "Flame of New Orleans".

Tuesday morning was again spent in shopping and general sightseeing. The planes were grounded because of foreign material on the flight controls. So no ferry trip home. The Beechcraft brought two more ferry pilots; Swindle and Merritt, so after dinner we all went to another show and saw "Scarlet Pimpernel".

Wednesday, August 27, I had my first flight in the Curtis 81-A Tomahawk or China Fighter.

I took off from Rangoon at 10:30 and after flying thru several squalls and never over 1000 feet; usually at less than 200, I arrived back at Toungoo about 11:15 o'clock.

After lunch I again had a half hour flight during which I got familiar with the plane and had a merry time doing vertical slow rolls and flying on my back.

A distribution of the cake and candy I brought back brings me to dinner time.

We started flying in earnest with three or four planes in commission with as many chutes. For the past week averaged almost two hours each per day, most of the time spent in individual combat, going up above the overcast thru any small hole which could be found in the vicinity, and taking on any single, pair or formation of ships. Many times two planes would start combat and within five minutes there would be four or five planes winging in and out of each other's way. It sure kept one's eyes open to keep all planes in view and still trying to get an advantage on one. By the time you would be thru fighting and get ready to land, you would have to acquaint yourself and try to figure out where the field was. Usually if you could find the railroad and Burma Road, you could figure out where the field was.

Robert's Plane was number 85.
(Copyright Charles Baisden)

Authors Note: Chennault provided training for all his pilots. He called his training "Kindergarten." The pilots were required to log "60 hours of flight training and 72 hours of lectures."[18] Japanese flying tactics were taught and the AVG pilots were instructed how to attack based on the type of plane they were flying and the way in which the Japanese fighters attacked. He instructed the pilots to "fly in pairs – stick together.....enter combat with the altitude advantage...dive, hit, and run."[19] Chennault told the men, "[Japanese] pilots have been drilled for hundreds of hours to fly in precise formation.......break up their formations and make them fight according to our style."[20] All the training provided helped

save many of the pilot's lives and Chennault's tactics were proven to succeed as the AVG downed plane after plane.

September 1941

This last period also saw September 2, come and go, which was none other than my 24th birthday – I felt magnanimous enough to buy drinks for the fellows who came from Mitchel – all in return sang the Birthday Song at dinner.

Besides flying everything is about the same except an earlier schedule for breakfast and work in the morning – 6:30 breakfast.

Today, Saturday September 6th, we had another insight into the industries of Burma – Teakwood. We traveled about 11 miles out of Toungoo in one of their famous dilapidated busses with hard seats and wooden windows. The elephants they use for movement of the trees, work only in the morning, but through a company official we had it arranged so that one elephant would be available in the afternoon to put on a show. The trip to the farm was thru the city and then into the most dense jungle growth I've ever seen. One could not see beyond the first foliage group. Not only did the many trees block the view, but wires would grove between the trees and shrubs making a curtain of Greene foliage. The road was of dirt, one way, and wound up and around a climb of only about 500 feet, but because of the dense growth, made it long and treacherous. Many were scared going around curves with a drop into impenetrable jungle on the side. One noticeable feature was the absence of flowers and colorful plants – everything was Greene – trees, vines, shrubs, plants and grass. We did see a small plant like an ordinary pot shrub, which would close its leaves and droop whenever it was touched or brushed against.

We had a little difficulty in finding the exact spot where we would find the elephant, mostly thru language difficulties, but as we were driving along, we suddenly noticed the elephant just inside the foliage, next to the road. We called for the elephant boy who lived in his small bamboo thatched hut along the edge of the road. He went to where the elephant was seen and called him out. He then proceeded to make the elephant kneel down, sit down and do various maneuvers. He then rode him to this house where he had a scaffold under which the elephant stood while he placed mats on his back and strapped on his harness for chains in order for the elephant to drag logs. Two interesting maneuvers these two went thru was that the boy dropped a loop in front of the elephant which he put his trunk thru then one foot at a time so that the boy had a rope around the elephant which held the mats down. He also had a loop of bamboo about 12" in diameter which he hooked over the elephant tail while he held his tail out.

The boy then took the elephant to a siding where several logs were already placed and had the elephant pull a log about three feet in diameter and 20 feet long. He also picked up smaller logs with his trunk from the road up on a bank about six feet high. He also thru*(sic)* a smaller log along the road.

Several of us had our pictures taken alongside of the elephant and one fellow ventured on top of the elephant and had a short ride. We passed the hat around for the elephant boy who had put on a good performance with his well-behaved beast.

A seemingly long ride on the hard seats brought us back to our little community.

Monday, September 8, was a real black Monday. While we were sitting at the luncheon table with empty dining room, we heard that two planes crashed in mid-air. After a process of elimination, it proved to be Bright and Armstrong, both Navy men. By one o'clock Bright had been located hanging from a tree in his parachute, unscathed, but mildly shocked. At first Armstrong's plane was not found but soon it was located almost buried in the jungle, both planes within five miles of the filed. Armstrong never did know what hit him.

Tuesday morning we had our first formation flight. We had five ships starting out in a three ship element and then followed by a two ship element, cross over, turns, string, back into two ship elements, three ship elements and practiced a Lufbery circle. There was no flying in the afternoon due to the services and burial of Armstrong. The services were held in the Officer's Mess Hall with all members present. From there we all drove into town and had the burial services at an English Cemetery. Chaplain Frillman officiated at both services. There were a fine bunch of flowers at the grave and a most impressive ceremony – a bugle call given by one of the Burmese soldiers.

An episode of interest happened today, Wednesday. We started out in a four ship flight of two, two ship elements to visit the alternate airport north of here about 125 miles. We did not quite get far enough north due to unpredicted drift, so buzzed a boat on the Irrawaddy and proceeded back home. The drift again took us to our left so we passed the field and after a while, we all became a little anxious as to our position. We then were far enough south to hit Rangoon, but not knowing our exact whereabouts, the formation broke up into every man for himself. I thereupon headed due north, following the railroad and Burma Road. After running on gas tank completely empty and just running dry on my second, after 25 minutes of flying due north, I saw the field and switched over to my last tank of 30 gallons. The surprising feature was that as I made my traffic pattern to land, I saw all three of the other planes, so my wonder as to their predicament was brought to a good end. We all landed with approximately 20 gallons of gas after 2 ½ hours of flying.

The Third contingent of pilots and mechanics arrived on Monday, September 15, almost a month to the day after us. They left San Francisco on July 22, went to Honolulu, Sydney, Australia, Manila, Batavia, Singapore, Rangoon and here, so they had a longer and more diversified trip than we. They were 53 days at sea, or on their way. There were 16 pilots, 6 Navy and the rest Army. The others were Mechanics and armament men. Those already here waited up for them to come in about 10 p.m. and have a few drinks with old friends. I knew two who were with me in Flying School. Tuesday they looked around and did a little shopping in Toungoo. Wednesday they started checking out – one got a prop when he used the brakes too energetically. Wednesday night we also had the first of our movie pictures here on the field. We are to have our own projector and 12 pictures a month coming up from Rangoon. They will be American pictures – just a few months old. I saw my first one Friday night. It was "Lillian Russell" – rather entertaining!

Saturday morning our long awaited visitors came in. We were expecting several English visitors for a week and spent three days practicing a review formation for their benefit. They arrived about 9 o'clock in a Blenheim, Lockheed Hudson, and our Lockheed '12. There was Sir Robert Brooke-Popham, a General, and several other ranking British Air Ministry Officers. We flew an 18 ship formation, first came over in three ship veer off the three squadrons of two elements each. The second pass the first squadron was in a two ship element; the second in company front, and our third was in string. It turned out fairly well. The visitors left about 11 o'clock and everything resumed normally.

Monday morning proved to be another "Black Monday" with poor ceiling rain all around, low clouds and a thorough day. Death came again. Maax Hammer was listed as missing after he did not land at 11 o'clock; there was no news of him, so at 3 o'clock with poor conditions, ten ships were sent out as a searching party, but about 5 o'clock a call came in that a plane was seen to have crashed about seven miles from Toungoo and the pilot was killed. Several cars departed with the aid of several native guides, a tough trek thru jungle brought the crashed plane into view. It was believed that Hammer in an inverted spin, could not pull out with such a low ceiling. The memorial services were postponed from Tuesday till Wednesday afternoon as Rev. Frillman was down at Rangoon and had to come up. There were memorial services in the Officers Mess and at the cemetery *(sic)* as was for Armstrong,. Maax was with me in primary at St. Louis and at Randolph and Kelly.

Monday morning also brought the Air Vice-Marshall of Far East Air Force of England to our home base. Due to bad weather that afternoon, Tuesday and Wednesday (a late monsoon period) we did not fly a review for them till Thursday morning. It was the best yet.

I was also relieved as S-4 (Supply) and made Assistant Operations Engineering Officer.

Mr. Alsop, a newspaper journalist came up here with our group to act as a publicity man later on. He was formerly in Washington doing political writing.

Another newspaper man who just stopped here on his way into China was Mr. Leland Stowe, of the *Chicago Daily News*, war correspondent staff. He has been in Norway, Finland, England,

Spain, etc. and in the Far East where he saw some real action. He stayed here for two days and gave us a short talk on his experiences and view on the war situation.

_____

October 1941

Flew down in the Beechcraft on Thursday, October 2, supposedly to get a P-40 to ferry back, but there were only two ready, so only two of the four of us could fly back; Moss and I were left out so we came back on the Beechcraft in the afternoon without even leaving the airport. The trip was therefore a failure as far as I was concerned.

The week from the 6th went by rather slowly due to Wednesday being maintenance day and there being no flying. Also had a couple of days of rain which further held up flying. Thursday I was in the Airdrome office, spending my time in the tower watching everything from the outside looking in.

We were also honored with a visit of several ranking officers from the U.S. Army mission on their way to China – General Scott was among those present. At the same time, there were two air Corps members of the Chinese American attaches, Major Rosendorf was one, the other was a Captain.

The Major incidentally was once with the 8th Pursuit Group when it only consisted of the 33rd Squadron; back in 1932.

Saturday a group of us went into Toungoo to see a traveling show. There were a large group of small open stores selling mostly foods – open air restaurants. The natives would be squatted on the ground around the large tray of many varieties of food and eating as if they were alone in a palace. The main tent was of bamboo. It was about 100 x 200 and sloping roof of matted bamboo held up by a few bamboo poles.

The whole area was covered with small bamboo mats about 3'x5' with a larger numeral in one corner. We finally found out that those were the reserved seats! The patrons brought a rug, pillows, and thermos jug and probably food. They would come at 7 p.m. and stay till 7 a.m., the next morning, watching a continuous performance. The first thing to catch our attention was the orchestra.

The main piece of music was a circular row of skin-metal drums, ranging from 2" in diameter to about 10". There were about 20 of them and the musician; squatting on a table in the center, would twist around tapping the drums with his fingers; sometimes hitting two or three with one hand, his fingers being outstretched enough to strike several small drums. There was also a similar cage with cymbal instead of drums. There was also a regular drummer, with three or four large tom-toms; a symbol player, two squeaky horn players, and the oddest of all were three men who played a split bamboo instrument. These instruments were about four feet long and four inches in diameter. They were split lengthwise and tied together at one end; the musicians would pull apart the opposite end and let go and a resonant "clap" would be heard. The whole orchestra playing together was a weird sounding affair, the melody and rhythm being hard to follow as each man seemed to play at will.

The stage was of the usual size with appropriate Burmese scenes depicted on the curtains. The first scenes consisted of a group of Burmese girls on one side and men on the other, kneeling, facing a statue of Buddha and they sang several songs.

Then followed several scenes of the girls dancing and singing. After that followed two acts of comediennes which we could not appreciate. We seemed to have seen enough by this time, so after a peek behind the stage where we saw the space divided into a sort of dressing room, we left.

From the "Opera" we went to the railroad station to welcome the latest newcomers into our ranks. We said "Hello" to most of them, had a glass of tomato juice and left for home.

Sunday, the 12th, was sightseeing day. Doc Priso got a station wagon and seven of us piled in with a box of canned and bottled goods and took off about 10:30 a.m. We drove out of Toungoo, east into the mountains. After 30 miles of hairpin curves, narrow roads, and plenty of dust, we reached the summit of that particular mountain, 4368 feet above sea level. Because of the cloudiness, we were at times looking down on top of the clouds, stagnant in the valleys. We stopped there at the P.W.D. house and hungry consumed our "canned" dinner. It was very tasty at that! From there we drove a little further to a tea mill. There we saw where the tea is laid out to dry in a constant temperature room (85 deg. F.), washed and rolled, left on a cement floor to ferment for about six hours; then dried in ovens; separated in size by machines, hand cleaned, sifted once more and then packaged. Some of the tea after sifting is pounded by a very primitive method of two women stepping off and on a fulcrum pile driver while a third woman took care of the hammer and tea. The tea would be pounded into a very fine dust which would dissolve almost immediately. There were four grades of tea depending upon size of the leaves. The English manager took us on a tour of inspection thru the three story wooden mill.

On the way back home, we were met by a native servant along the road who directed us to Mrs. Cote, M.D., M.B.E. This woman is about 89 years old and has lived in Toungoo for the last 20 years. She was in Rangoon practicing medicine for 30 years previously. She originally came from Canada. Her wit and speech is of the finest; knowing all the latest word information and topics of interest. Her Indian maid, who speaks good English, made us donuts and tea which was very enjoyable, not having any such delicacies for some time. The M.B.E. is an honorary title given her by Britain. Member of British Empire.

Bob sightseeing.
(Courtesy Virginia S. Davis)

And so back down the winding cloud hidden road to Kyedaw Airport, supper and a show – "It's a Date" with Deanna Durbin.

Sunday also proved to be a surprise in that upon returning from our little journey, I found a letter awaiting me from home. The first letter since leaving home. Sure was welcome.

Thursday, October 16, we were again honored by a visit of English officials. This time it was the Right Honorable and Mrs. Duff Cooper. They came up in the Lockheed and as soon as they landed and got out of their plane, 18 of us, six from each squadron took off and passed twice in review. The first time we were going north. The three squadrons were in elements of V's. The second pass was going south. The 1st Squadron was in company front, we were second in string, and last was the 1st Squadron in three elements of twos in V. It seemed a very good review from the air and were told it looked fine from the ground. The visitors were only here for several hours making a short tour of inspection of the airdrome.

Colonel Hoyte, now a member of the American Observing Corps in China, formerly a commanding office of several pursuit groups back in the states, was here for several days and also gave us a little talk on cooperation and teamwork.

Due to Bob Walroth's leaving, I was made Engineering Officer temporarily. We had a good scare Saturday the 18th when, at about 4 o'clock two foreign looking planes flew overhead and we could not make out their type. We were of the opinion of their being Japanese observation planes, but found out Sunday that they were the new British training planes. For this reason and because of the new crisis of the Japanese War Cabinet collapse and reforming, we started arming our planes and having several fully loaded and ready to go. We are to start an alert set up – just in case.

By the following Tuesday, we had our first alert staff working. There were a total of 12 planes, four from each squadron. All of these were fully armed and loaded. Each day a new squadron would furnish six pilots and reliefs for these six planes. The other six planes would be taken by the first pilot to get it.

Our Squadron #3, had its first taste on the next Thursday. We awoke at 4:45, got a truck at the pilot's mess and were ready to take off at 5:30, if necessary.

The mechanics arrived at about 4:45 to check and warm the planes up. We were brought hot coffee and sandwiches about 6 o'clock and then went for breakfast at 7:30. That day we had maneuvers with three Blenheim bombers. The alarm was sounded about 9:30 that the rats had left Rangoon. Our six alert ships took off along with 12 others. Three groups of six each, were sent out to intercept them. About 10 minutes out, I sighted the three rats about 10 miles to our left and 3,000 feet lower. I called the flight leader, Olson, and said "Three rats at 3 o'clock." I then caught up to him and wobbled my wings to get his attention. As soon as I got it, we started in their direction, but by then I had lost sight of them. We continued on the course for several minutes and then saw a bi-motor ship about 6000 feet above and to our right. We quickly climbed, but soon saw it was the CNAC [Chinese National Aviation Corporation] Douglas DC-2. By then we lost all track of the real enemy and started down to get ready to land when we got another report of the position of the enemy so we took off again in pursuit of them, but soon received the report that the field had been reached by the imaginary enemy and blown to bits. So there was nothing for us to do but go back and land. None of the other squadrons had not seen anything of the enemy so I lost my chair of being "here of the day" and instead became goat by losing sight of the enemy. Nothing much was said except a little ribbing by some of the fellows. It was a lot of fun even if we did not prove successful in our mission. At least I learned something.

Our squadron also had its first major plane catastrophe on the same day after the maneuvers when Hodges ground-looped and tore off both wheels and skidded on the belly onto the grass alongside the runway. He was not hurt at all, but the plane was a washout.

A fancy dress ball in aid or war charities, with Major General J. Bruce Scott, M.C., at the railway institute on Friday, October 24, was the event of interest for the week. To prove its worthiness, I put on my tux for the occasion. There were quite a few people in costume, the women mostly trying to represent some phase or form of peace, even to one with angel wings. About half of the people were Burmese so I had an opportunity of dancing with several of the locale belles. Of course, I was particular and fortunate enough to dance with only the prettiest of the girls. With several drinks to hasten and enlighten the frivolity, the evening was most enjoyable and was almost sorry to leave at 1 o'clock except for being slightly tuckered out after so new an evening of entertainment.

Black Saturday, 25th of October. They say accidents happen in threes – well, we had our third mortal catastrophe and I hope our last. While acting as relief at the alert station about 9 o'clock, Pete Atkinson took off on a test hop. About 15 minutes later, we heard a terrific whining sound and glanced up into the blue sky to only see an airplane disintegrating into a million pieces. The whine continued for a few seconds as the pieces of the plane floated down while what looked like only the engine section plunged onward at a terrific speed – another second of quiet followed by a loud roar as if a bomb had exploded as the motor hit the ground. A wing continued floating down and other small pieces tumbled down among them, as we later found out, was Pete still strapped in the seat, plummeting earthward to crash thru a tin roof of a house and unnecessarily complete his final landing. Another friend and good pilot had landed his last plane.

From other persons observation it appeared that the plane was diving at a great speed, the prop over running considerably, making as loud a whine as anyone ever heard; the plane started to pull out of the dive and the plane started disintegrating; the tail section coming off, then the wing, then the whole plane coming apart like a fire cracker shot off in the air. The almost indestructible diving plane just gave out at the wrong moment and took away another test pilot.

There was no flying scheduled besides test hops, but all flying was called off and all planes grounded till the tail assembly and wings were inspected.

The following day we held the last rites for Pete. At the Pilot's mess hall, we had 24 pilots and crew chiefs form a honorary pall-bearer double line thru with six of us pilots from the old 33rd acted as pall bearers. The services were given by Paul Frillman and after the very impressive ceremony at which almost every man attended; we formed a long procession of seven cars, three station wagons and two buses into Toungoo to the cemetery. There we again had the double line and at the grave alongside the other two concrete slabs of former pilots, services were again given. This time by a Catholic priest as Pete was a rather devoted Catholic and wanted it that way. So at 5 o'clock of the afternoon of October 26, 1941, a great friend and pilot passed from the level of this earth to a better one beyond.

For the next few days, there was quite a scramble getting the tail assemblies of all the planes inspected, which caused very little flying to be done.

Once again the English High Command paid a visit to our training base on Thursday, October 30. This time it was General Wavell and his staff of about three other Generals, including Gen. Scott of Toungoo. They were flying in a DC-2 accompanied by a Blenheim. They were here only a few hours, but long enough for us to hurriedly gather together 18 inspected planes and put on a review. Because of rush, it was not as good as might be expected, and also offered a little excitement when, while landing, Sandell got a wing tip but took off and circled the field and landed. Sawyer also had a little bad luck in blowing out a tire, but doing no damage to the plane otherwise. Moss also had the same thing happen just before the party arrived.

The previous evening another contingent of pilots and clerks arrived from the U.S. after 58 days out of Frisco. We got three pilots. They started flying Friday and since they all came from the

Navy and were flying big boats, they had some difficulty. Conant, one of our gang, landed 15 feet up in the air and when he did hit the ground, he ground looped, tearing up an airplane, but not doing any damage to himself.

November 1941

Sunday morning we went for a swim up in a rushing mountain stream about 20 miles up from Toungoo. There were eight of us in a station wagon. We brought sandwiches from the mess and bought some canned fruit and had several thermos jugs of an ice cold fruit drink. It took about half an hour to get up to the 500 foot altitude of our little swimming hole. The stream at this spot is about 50 feet wide and about 30 feet of it is over eight feet deep for a stretch of another 50 feet.

The current is almost too strong to make any headway against it, so it is quite a workout to just hold your own. After a swim and a little lunch, we continued up to Than dung and the tea factory on top of that mountain. On the way back, we stopped at Dr. Cotes for tea and donuts.

If Sunday was a day of peace and quiet, then Monday was a day of disaster and rumpus because with seven accidents, it could not be very peaceful. Among the affected was Conant, who after cracking up on his first flight, nosed one up after overshooting on his second flight this date. Another rather avoidable one was caused when a mechanic rode a bicycle into an aileron and damaged it enough to be replaced. The third casualty to our squadron was when McAllister nosed one up tailing to the line. One of our new pilots left his prop switch on maneuvers, dove over, ruined the prop and burned out the bearings. Then in the afternoon a mechanic of the 1st Squadron taxied one plane into another damaging one pretty badly and putting the other out of commission for several days. Then Squadron Commander Sandell got his third ship – he nosed one up off the runway. Those are the seven and one minor one – a flat tail wheel on Sawyer's plane at Lashio on a cross country finished the day of woe.

They came thru with several more ships Tuesday and on Wednesday Conant got his third one – he blew a tire on a touch and go landing and held it straight for a while, but a landing gear gave way and he ground looped right towards a ship being bore sighted with about six mechanics working on it. He stopped about 20 feet from it. *(sic)* This plane was really washed out. All we got Thursday was a blown out on a ship taxiing so not much damage done except that we are running out of spare tires which we are getting from wrecked planes – also no spare props.

On Thursday an order came out of naming Squadron C.O. vice C.O. Flight Leaders and Wingers – I was named A.F.L.!!! Should mean a raise in pay I hope!

That Saturday, three Blenheims and six Brewster Buffaloes flew up on their way to Hayhoe. The pilots were those we met at Singapore and now they were transferred up here. They remained here only several hours, gassing up and having lunch. They flew over the field later in the afternoon on their way back.

Tuesday I was incapacitated with a slight touch of dysentery, so I spent the day in bed and did not fly the next.

Thursday got in an hour flying for a change and Friday I got my first gunnery practice. Between the iron sight and half the guns not firing, I did not do so good. The rest of the morning I spent as Airdrome officer in the control tower. Also test a ship with vanes on the dust pan of the wheels to try to get them spinning before landing, but the experiment was not very successful.

On Sunday morning I received a post card from Grace and Gus from Albuquerque mailed September 18th. Also took a bicycle ride to a nearby pagoda to take a few pictures. It is a small compound, a

Robert sightseeing.
(Courtesy Virginia S. Davis)

stone shrine and several small stone shrines – one building with a golden Buddha but not many jewels on it. Another alabaster statue and several old objects like drums and bells. On returning to the field, we watched the Beechcraft land with Gen. Magruder, one of the American Mission to China. He stayed overnight and caught the CNAC the next day.

One day the next week, two Brewsters came up and Shilling took one on in a dog fight and beat him three out of three. This morning, Friday, another group came up, four of them, and Bacon split one for on. It seems the Tomahawk can hold its own.

Last Thursday, Olson, Howard, Neal and Shilling flew up to Kunming to look the place over. They made it in about 2 ½ hours, averaged 245 mph. They returned today.

The next few weeks passed by like greased lightning. Nothing happened but the time went very quickly.

December, 1941

Monday, December 8th will probably prove to be an outstanding day in my aviation career for upon arriving at the line about 6:30, we were met with the astounding news that on December 7, Japan had attacked Hawaii, Philippines Islands, Wake and Cream Islands of the United States Territories. They attacked Hawaii early in the morning and without any opposition did considerable damage to the airfield, killing over 200 soldiers and about 3000 casualties were due to the bombing of the cities. There was a naval battle near Honolulu when Pearl Harbor was attacked and it was rumored that the Oklahoma and West Virginia, both capital ships, were sunk or badly damaged. Several Japanese ships were shot down in the melee. The attack upon the Philippines was very similar and here they tried to land parachute troops on several of the islands, Tuzon in particular. Both Nichols and Clark air fields were severely bombed and a good number of dead and casualties. The Clipper was shot at and burned. Here also a few Jap planes were brought down and a German pilot was found in one of them. Wake and Guam islands met similar fates.

The Japs also attacked the Malaya Straits, simultaneously bombing Singapore and Hong Kong. There were rumors of parachute troopers being landed above Singapore on Malaya Straits.

Thailand also felt the force of the Japanese army when it held off the attack for five hours, then resistance stopped, but again drove back the Japs.

It looked like a well planned Blitzkrieg for Japan. They seemed to have won the first round, but the fight is far from finished.

Here at the field, all the ships ready to go were put on the alert fully loaded, oxygen lined up and even kept warm. Nothing happened here at the field but a lot of post mortems and prophesizing went on. The Navy boys took quite a verbal beating for letting the Japs attack without holding them off or letting them thru.

Tuesday brought more reports repeating the previous happenings and a few confirmations came thru although many reports of new naval battle, sinking of ships, landing of parachute and land forces came over the radio from all parts of the world.

That afternoon about four o'clock after an otherwise calm day (besides getting up at 4:30), an observation plane was thought to have been near and immediately about 20 planes took off in search for it, but none located any foreign ships. A night alert group of six was put in effect and almost came in use. About 3 o'clock the next morning we were awakened by the shrieking of the siren, denoted an air raid – in a few minutes we heard the six ships taking off. By then we were dressed and outside our barracks next to the five bomb trenches constructed for such a purpose.

We could hear our own ship flying overhead, but even with a bright half-moon and a clear sky, we could not follow their course. I went to the field about 3:30 to await developments. By then most everyone had congregated there for the same reason as I with the thought of a bombing in sight. However, discounting several flashes thought to be flares and others thought to be bombs, nothing happened, and after a half hour the all clear signal was given and the runway was fairly lit by hard lanterns. Soon the planes were seen to turn on their wing lights and come in to land. The first two made it fine, but the third with Tex Hill overshot and went into the brush; working the plane out, but doing no personal damage. The other three ships landed in good order. Soon it was time for breakfast so had a not too hearty breakfast after that small alarm. The rest of the day passed as per usual, but for the tenseness in the air. The radio was listened to with interest and everyone expressed his views on the happenings and consequences. There was a report of Jap planes seen 20 miles off San Francisco. There were no further reports of the battles in the Pacific.

There were also no reports or confirmation of any bombings in Burma so we must have been mistaken in our little escapade which cost us one great P-40.

That afternoon, we also sent out a camera ship with two escorts over Thailand for a reconnaissance flight, but they returned with no important news. That day we also heard the rebroadcast of President Roosevelt's Fireside Chat, which all the fellows seemed to acknowledge. An alert crew has held that night but although there was no alarm, there was the telephone, tenseness and other precautionary noises to cope with for a none too pleasant night. *(sic)*

Thursday morning the reconnaissance flight took off for Jevoy and refueled and then flew over Bangkok at 26,000 feet for pictures. Upon returning and developing the pictures, 93 planes were counted on the field. Seems like they took Thailand over and moved right in. More news of the sinking of several British ships off Singapore. About 3 o'clock, I heard our squadron was to go down to Rangoon to help the R.A.F., but because of bad weather, could not take off. We did get off at 9:30 the next morning and 18 of us made the flight without any trouble. The truck convoy that left the previous evening with the ground crew was there waiting for us at Mingaladon.

A "Yank with the R.A.F." might now be my nickname. We got our planes dispersed, gassing facilitated, room and board taken care of and got ready for action. We were fortunate enough in not getting any the first date, but on Saturday after being kept up most of the night with a mild case of diarrhea, I had tea for breakfast, straightened out a few matters and returned to the barracks, but was awaken by the siren. There were two observing planes seen overhead, but except for several Brewsters giving chase, no other action was taken. Back to bed. At about 1 o'clock the shrill blast of the siren, the call of fellow pilots that they were coming, and I was out and dressed in no time flat. I missed the car and by the time I walked to the line, all the planes had taken off. The report was that there were 27 enemy planes coming. Within 15 minutes, there were 18 Tomahawks and about 16 Brewsters up and after them. Two of the 40's had to return because of prop trouble. This alarm again proved to be in vain because the enemy turned back about 100 miles from here. Hope they do it every time. We quickly serviced the planes just in case they returned, but not till about 9:30, just after I got to bed and about to fall asleep – thinking of some good air tactics when the shrill siren sounded and roused me from my semi-sleep – fumbled the mosquito netting, grabbed my clothes I had set next to my bed for just an emergency, threw them on quickly, grabbed my gas mask, helmet and gun and dashed out the door, down the stairs and next to a bomb trench where I was already beaten by several others. Soon it became very quiet except for someone calling, "Put out that light". After about a half hour of this tenseness, waiting for something to happen, the all-clear siren sounded and back to bed. This time I really slept.

My stomach still bothered me for the next few days, but except for a half-hour check hop on Tuesday, we did not do any flying, so I had no cause of worry about flying. We have a three shift alert now, of two groups, from 5:30 to 8:00; 8:00 to 1:00; and 1:00 to 5:30. This gives each fellow a chance to get a little sleep when he is on the 30 minute alert. The five minute alert remains on the line, keeping the planes warm and in readiness.

We did get a chance to go into Rangoon Sunday night after a rather hectic trip in. There were no taxies out at the field so five of us rode atop a gas truck into Insien where we took a local train into Rangoon and a bicycle rickshaw to the Savoy for supper. Took about 2 ½ hours from the time we let the barracks till we got to the Savoy. As my stomach was upset, I only had a brandy to try to calm it and a bowl of soup – at about 9:30, Haywood and I got a taxi and I came back out, not a too eventful trip.

At first I really thought I missed hearing the alarm and a Jap bomber had buzzed the field, but as I became more fully awake from my afternoon resting, and did not hear any bombs, I thought it was a formation of Brewsters. Someone on the porch said a consolidated B-24A had just passed over, so I quickly ran out and saw it disappear below the trees preparing to land. Several of us quickly dressed and ran out to the runway just in time to see it turn around and taxi back to the line. We took one of our little Stud. "Champions" and drove down to the plane. With high hopes and feeling elated over meeting the American crew, but only one fellow was known to any of us. He was the navigator in Hodges class and later I found out he knew Chuck Beves very well. They had left

Bolling Field and proceeded via Miami, Puerto Rico, Brazil, over the South Atlantic to Africa and up to Cairo, Egypt, where they picked up General Brett. The plane stayed overnight and then took off for Calcutta for safety. Most of the officers, including the General remained here.

The next morning, Thursday, I went up on my first patrol. The British were expecting an important convoy and wanted some fighter protection. We only had four ships up at one time; two at 10,000 and two at 15,000. We were first at 15,000 for an hour then at 10,000 for the second hour. We only had a 30 mile strip to patrol; about 20 miles East of Rangoon between the coast and Thongwa. Nothing happened during the tour, except a sore fanny from the ride.

Another patrol Friday at 17,000 feet for a couple of hours, but nothing sighted. We have the alert now from dawn to dusk. Same schedule Saturday.

Since I was to have Sunday off, I went into town Saturday evening. Several of us went to see "Hold that Boat" with Abbott and Costello, which was quite a scream. Then we went to the Savoy for a steak dinner and a few drinks. From there I went to the Silver Grill, where I met Kennedy with his girl, which called for a little Bird-dogging, so I had two or three dances with her – the first real dancing since Los Angeles, except for the Dress Ball up at Toungoo. I left for home rather early – Kennedy had the CAMCO station wagon, so he was kind enough to take me home. A rather pleasant evening was had. I had a notion to go into town Sunday morning to spend the day there, but was unable to get a taxi, so Hedman and I didn't go in till after luncheon. We went to see "Mr. Jordan Returns", then had tea at Savoy and we went to a show again. I saw "Bittersweet" and he saw Abbot and Costello. Afterwards we met at the Savor for another spot of tea where we met Overend and Older. We all came about 10 p.m. Monday passed as usual.

The 3rd Pursuit Squadron *(Courtesy of the San Diego Air & Space Museum)*

BUT – Tuesday, (12-23-41) Now, that was the day – war came to Rangoon in the form of 38 bombers and 27 pursuit.

We received an alarm about 10 o'clock so six of us took off for 17,000 feet. Dupouy had three and then Martin and I and Jernstedt the last three. At about 10:30 we were at 17,000 when I saw 21 bombers in a beautiful tight V of V's, coming in towards Rangoon at about our altitude about 10 miles away. We climbed about 2000 feet and then dove on them in a string. I aimed at the leading formation and from a front quarter diving attack, was able to sweep the whole formation with no apparent results. After diving down and away, I turned around and got more altitude from which I again followed Martin in a rear quarter diving attack. I made the mistake of again sweeping a formation instead of picking out one ship. No results again. Since we had broken formation by then, I started looking for someone to hook onto. I thought I saw a group of about six planes which I thought were another of our group, so I started toward them to tack on. Upon approaching closer I saw the planes had fixed landing gear. I thought this odd and upon getting closer I saw the large red circle on the upper wing starring me in the face. I quickly turned and gave it full guns as I saw two planes take after me. They only followed me up a few thousand feet and then dropped back down into formation. I kept climbing up to 20,000 feet where I looked around for someone to hook on. I couldn't find anyone so after a half hour, I went down and landed. Although the bombing was terrible and devastating, but it was a picturesque sight to see the large bombing formations of 21 and then another of 18 that came in a second wave. They held a beautiful close formation and you could see the black particles of smoke on their bombs hit Rangoon. The spots were scattered and about 10 or 12 fires were started, bellowing smoke straight up and forming a mushroom at about 3000 feet. There were also many bombs dropped along the docks but it seemed as no damage was done – only splashes of water as the bombs hit. Around the formation were black spots as the anti-aircraft were being fired upon the bombers. Out towards the airport I could see several fires as petrol dumps were hit – one in particular was giving off a black tower of smoke which rose perfectly vertically and formed a head about 3000 feet. Spots were also seen on the runway where small craters were made by dropping bombs. Throughout my flight I could not see any pursuits, but upon landing I heard the other fellows' stories. Most of them seemed to be by themselves in their attacks – getting in close enough to see their hits go home. There were about six fellows claiming hits. We were missing three planes but after about 20 minutes, Greene came back in a car rather banged up. His plane had been badly shot and he bailed out. His chute was loose and jarred him and scratched his neck and face. He was stiff all over. On the way down he was strafed but luckily enough not hit. Someone had seen Martin hit and dive in, but no confirmation. Gilbert's plane was found with him shot and burned – not much left of him. So we lost three ships from being fired upon and crashed and two others shot up too badly to be serviceable. Reed upon landing in an afternoon patrol, hit a bomb crater damaging a wheel and wing; which leaves us with 12 serviceable ships. By the by – Greene was using my plane when he got hit, so I feel sorry about that, but happy to know my chute worked!

While one squadron of bombers hit Rangoon, another bombed the field putting many direct hits upon hangars, petrol dumps, operations buildings and the runways. The pursuit ship came down and strafed the field. They did a great deal of damage, burning up grounded planes and ruining many buildings.

On the humorous side was the fact that most of the bearers and cooks and servants left the field and never returned. We had to make our beds, get our own food from the kitchen and mostly we had cafeteria, buffet style of serving. The rest of the afternoon, we kept up at least six planes on patrol, but nothing was sighted.

There were eight bombers and one fighter confirmed, but there were probably more that did not get back to their bases. I think I put enough lead in some of them to sink them.

My reactions throughout the day were varied. When we first sighted the enemy I marveled at the beauty of the formation, and as I charged my guns and turned on the switch, diving down on them for the first time, I had a slight feeling of exultation and mostly wonderment – wondering if I would bring anything down. The next pass I made as casual as on a target ship, not even worried very much about return fire. But as I passed, I really stuck the nose down expecting to be hit any second. My greatest scare came when the two Jap pursuit ships started after me, but after I lost them, I just kept my eyes rolling like a (expletive), looking for enemy. I think I'll do better hereafter using more discretion and less caution.

After just retiring that night, a false alarm came through which caused us the unpleasant necessity of going out to a trench.

Wednesday brought no news of Martin or more than nine confirmations. We kept up a patrol all morning with six ship flights – around noon a report of nine enemy planes came thru – everyone took off, but nothing was sighted. We patrolled all day till dusk, putting in five to seven hours each. The mess facilities were not much better, but we did get some food.

"CHRISTMAS IN THE TRENCHES" or more appropriately, "War in the Air". The morning started out pretty calm, even though the Japs promised us over the radio that they would give the Americans and Rangoon a real Christmas present. About 10:30 we got a call to scramble to 15 angels – I was on Bishop's wing in Dupouy's flight. We took off first, but Bishop did not wait for the rest of the flight. We got up to 18 angels and cruised around for a half hour, then not hearing anything on the radio, Bishop and I dove down over the field. We just about got to 2000 feet when I heard the radio say not to pancake. We started climbing again and as I got about 4000 feet, I looked around and saw a burst and then bombs exploding on the field. We must have been directly under the bombers and did not see them. I expected any moment to hear some ships diving on us as we were slowly climbing. I quickly gave it the gate and climbed as fast as possible. I got to 18,000 feet unmolested and I then saw a tight 30 ship Jap bomber formation going southeast over Rangoon at about 20 angels. I started towards them and as I got closer I noticed a single Jap Navy #96 pursuit ship, with fixed landing gear. I had about a 1000 feet height advantage off to his left, so I turned into him on his rear and dove down below his rear and climbed up his blind spot under his tail. As I closed into within 50 yards, I opened fire with all six guns. I could see the bullets entering his ship and he suddenly made a very quick wing over to the right, and I had to turn the opposite direction to miss him. I turned around to see if I could find him, but he was probably going down too fast. I didn't follow him down so cannot say exactly where he fell, but from my close fire, he could not have possibly escape destruction. I dove down for speed and then climbed up to 16 angels – I could not see any planes so I came in to land. Pending confirmation, I think I have my first plane.

The other fellows did very well – Hedman accounting for four, McMillan got three and was shot down himself. We thought he was a goner when he did not return, but after supper about 8:30, he walked in the mess hall with a bandaged arm and hand. He made a crash landing in a rice paddy and finally got the natives to help him. He got a horse and rode about 10 miles to Toungoo where the police took him to Rangoon on a police boat and then in a police car up to the field. He had a Jap Saber which he got from a bomber shot down on the first raid. We sure were glad to see him walk in. Overend was also reported shot down and missing, so we did not expect to see him anymore, but at about 9:30 that night, we got a telegram from a small village that Overend was safe and we should pick him up at Rangoon about 10 p.m. He was not hurt at all, but shot down while shooting a bomber down. He also had a crash landing.

Our unofficial account was 10 fighters and nine bombers. The R.A.F. got six planes and lost four. This was their first loss in personnel.

The two ships shot down and Smith's and Hedman's were pretty badly shot up. Dupouy shot down one model "0" and in getting away, hit his wing tip against the wing of another causing the enemy to spin in and Parker came in alright with his right wing tips off and half of his aileron missing. Another McMillan feat. Older and Haywood stuck together and counted four together. Some of the fellows thought they saw some Me 109's but not positive.

The damage to the field was about the same as before with larger holes caused by bigger bombs. Only one Petrol dumb was hit and two casualties on the field. The greatest damage to the personnel was via food. We were very low on food and, of course, were without any kind of Christmas Dinner. In the afternoon, Mr. Bill Pawley came out with a car full of canned food and drinks which was really welcomed. So another day under fire passed with experience gained and ships lost. A sidelight of interest was that just after the air raid sounded, the British General – Wavell D.C. – 2 landed with her and General Brett of the U.S. Air Corps. They both had to take shelter in a trench and saw the raid first hand.

Friday morning we had an alarm and sent up about seven ships on patrol. We were up at 22 angels for two hours, but nothing was noted.

We expected them back Friday, but except for the alarm none came. There was a rumor circulating around town that there were pamphlets dropped over Rangoon that the Japs would drop 3000 parachute troops that night, so we all went to bed ready to pull a strategic retreat in our cars just in case. However, except for a very slight sleep, we were spared the wild ride that would ensue.

The worst part about Saturday was the utter lack of digestible food. Breakfast consisted of bread and tea – dinner was salmon, apricots and coffee – supper was bread and cheese with tea. The cooks had all gone and no food seemed to be in sight – somebody slipped up some place. For safety's sake we took six ships over to Satellite Field at Regu at dusk.

While we were in readiness at the alert tent during the afternoon, we saw two large impressive looking cars with a flag flying from the radiator cap. They came to us and saw it was the Governor of Burma, Sir Dorman Smith, with General Brett and several other high British officers. The Governor expressed his best wishes and thanks for our cooperation and the General gave his opinion as of the air service. It helped the morale of the troops!

Sunday we heard that we were to be relieved by another squadron from Kunming, so we felt a little relief. There were no alarms or patrols for us so we got a little rest. About 10 o'clock we were surprised to see one of our CAMCO trucks pull up in front of the tent with the mechanics of "B" Squadron. There were three pilots in with them so we got some of the news from up North. We found out the official dope on the 3 CW 21's which got lost between Lashio and Kunming. Mangleburg crashed on a forced landing and was burned. Shilling and Merritt crashed their ships, but did not get hurt. Since there was not any missing facilities, the fellows took a train back to Toungoo where they had dinner at the railway station and then took a train back to arrive at Mingaladon again about 10 o'clock. We did not have any alarms or patrols again, but were packed ready to go whenever the other squadron would arrive, which it did not that day, except for Newkirk and Bacon who came ahead to get things in shape.

Finally on Tuesday morning, 15 more of "B" Squadron planes came in so after a couple hours of hot-air blasting, we took off for Toungoo. McMillan had engine trouble so did not get off, so only 10 of us got up to Toungoo about 1:30. Had the first good meal in about a week.

The next morning we had a breakfast of eggs – the first in one week – got weather reports from up country and got ready to take off at about 10:30. I just got off the ground when I noticed oil on my windshield. My oil cap was loose and oil was frothing out. I came right back in to have it taken care of. The others saw I was O.K. and went on their course. I was fixed up in about 15 minutes, so I took off. I thought I might catch them so I gave it plenty of mercury, in fact I had a hard time checking my check points because of my speed. However, I hit Mandalay all right and headed for Lashio. Just as I got over the field, I saw that about six were on the ground and a couple more in their traffic circle, so I let my wheels down and started coming in. As I was just on my last leg, I saw Smith, who was top cover man, coming down so I beat him in for a perfect timing flight. The country up to Mandalay was flat and had lots of towns and roads, but from there to Lashio was a little rougher. There weren't very many check points and the course was over a very narrow valley, with rough mountains on either side.

At Lashio we were taken into town for a bite to eat while our ships were being gassed. We took off about 2:00 with a bad broken overcast. About 10 miles out we went through a large cumulus cloud for several minutes and when we came out, we were all scattered, but finally rejoined. The country over which we flew was perhaps the most beautiful and most dangerous I have ever been over. We were at 15,000 feet and every now and then only five or six thousand above the top of some rugged peaks. There seemed to be sheer drops of several thousands of feet between cliffs – no landing places for miles on end. As it was fairly well overcast, we could not see very far, except down, and that was not all enlightening. The strange thing was that in the remotest part of nowhere on the ride, top, or valley of a range would be a green patch of vegetation of some farmer. How they ever knew what was even 50 miles away was hard to conceive. There were places about eight to ten thousand feet completely covered with clouds which looked like a snow covered lake between two mountains. Large billowy cumulus clouds started in a valley and rose about peaks like smoke rising above pipe. Besides being somewhat jittery about the flight path, it was rather cold and it was a great sigh of relief when the lakes at Kunming came into view. We dropped down, circled the large grass field with a runway being built through the center, and then landed 6500 feet above sea level. It was the first time, besides Lashio at 2500 that I landed over two or three hundred feet above sea level.

We were rushed to a ready shack where it was warm and were brought tea and cakes, and then some winter flying clothes to keep us warm. About an hour later, we were driving from the field to our hotel on the other side of Kunming, thru the bomb blasted streets, with a milling throng of heavily dressed Chinese, out on the streets with their wares displayed for sale. Others carrying produce in baskets hung from a bamboo pole across their shoulders. I was surprised by the great number of American cars each trying to out blow their horns.

Our hostel consists of three narrow buildings, three stories high, of white or grey brick with red tile roofs of Chinese design with painted ends and carvings all along the edges and top. The Administration building is in the center with rooming houses on either flank. The enlisted men sleep on the south and we on the north. My room is on the first floor, of white calcimine, about 20x20x12 with a canvas bed with a soft quilt mattress, a dresser, locker and writing table – two windows for fresh air and light, and a small clay charcoal burner for heat, and not too much of that.

The service is very good and the food almost equals that on the good old Jagers frontier!

Authors Note: The end of 1941 proved the American Volunteer Group could fight and beat the Japanese. Chinese and American newspapers announced the victories against the Japanese. These announcements gave the people of the United States heroes to honor as the World War was progressing against the Japanese in the Pacific.

The Chinese Newspapers "heralded their victories and dubbed them 'Fei Hu,' the Flying Tigers." This was an important designation because since the 1911 founding of the Chinese Republic, the Tiger was the national symbol. The name, Flying Tigers, was the highest accolade that could be bestowed on the American Volunteer Group by the Chinese Government.

During the last days of December heard the exploits of the American Volunteer Group were broadcast on many U.S. radio stations and through these broadcasts, the Flying Tigers became national heroes. They were known as the most dedicated and hardworking fighters to serve anywhere. The Flying Tigers became heroes and legends at a time when Americans needed that most.

Robert Brouk © *R.T. Smith via Brad Smith*

January 1942

The first day here, incidentally, January 1, 1942, we slept till about 7:30. Had three eggs for breakfast, then drove into town to exchange some money and send a radiogram home.

Most of the people in Kunming seem of the poorer class, wearing almost all a blue cotton quilt long dress with tennis shoes, and varied colored long stockings seen thru the split in the bottom of the dress. The women work as hard as men, carrying their babies, a la papoose, carrying baskets on their shoulder poles, have large bundles on their backs held by chest and head straps. I get the impression their work is terribly hard yet they endure it without any wailing. Men carrying in tremendous loads, looking as if they might not last another step, keep trudging on till they deliver that load to get another.

The shops along the brick streets are of mud bricks or wooden, one upon the next, with wooden doors, open during the day for an entrance, and closed at night for their home. Most all shop keepers live right in their shops. There are a few larger buildings and some on a more modernistic design, but there are also many just barely standing, after direct hits from bombs. Many a wall is standing with an interior.

The first week was spent in a most leisurely manner. We were having our planes checked and had no duties so we were on our own, getting up about 8 o'clock, taking plenty of time eating breakfast, then just sitting around the charcoal stove in the bar, playing Acey-Deucy or reading. The afternoons were usually spent getting flying equipment or just riding around getting familiar with the lay of the land. One afternoon several of us took a little side trip about 20 kilometers away up on top of one of the mountains overlooking the lake. There we saw the "Temple of 500 Faces". It gets its name from the fact that the main idol building, besides the five main statues in the center, are numerous life-like images of people. These small images, about three feet high are in full detail (plaster of Paris) with vivid coloring. They range from grotesque dwarfs to beautiful looking women – some seem to depict scenes – there were three walls covered with them from floor to ceiling, and there looked like more than 500 to me – closer to a thousand. They were like a Christmas display of brightly dressed and painted dolls.

The temple itself seemed to be built on a plateau of a cliff, rising 2000 feet straight from the lake. It gave a marvelous view of the lake, city and ranges of mountains surrounding the valley of Kunming. On the way back we stopped half way down the mountains to practice our gunnery with a 22 cal rifle and my .38.

Monday the 5th, we had the night alert of six, so I happened to be chosen and spent the afternoon getting details settled. We used mostly the 1st Squadron ships as ours were not finished as yet. We were on duty from 6 p.m. to 5:30 a.m. Nothing happened except I had a slight case of diarrhea which kept me awake most of the night, whereas we might have slept. I also was on the alert Wednesday night, but this time I slept better. Thursday afternoon, Sawyer, Greene, and I went in the Beechcraft to Yunnanyi for a three day vacation. Sawyer was told to take a rest and to take two fellows with him, so Greene and I were the fortunate associates. We left about 2 o'clock and started flying west along the Burma Road, only after about 30 miles, it divides and true to form, the pilot took the wrong fork and came to a field which he thought was it. We landed and found it was Shungshee about 20 miles northwest of our course and 50 miles from Yunnanyi. We gassed up and took off. After crossing several ranges, we came into view of a large valley with numerous lakes and small villages. There seemed to be about 40 lakes and as many villages. The field stood out like a sore thumb, especially the new stone runway coolies were building by hand. There are several hundred coolies just breaking large stones into little ones and placing each in place by hand. It is a sight to see – comparable only to the Burma Road construction gangs doing the same slow tedious work. They get paid 10-20 cents a day, while those on the road only get food from the Government.

We finally landed about four o'clock and were met by the three instructors stationed there along with Capt. Carney. We went to the house belonging to Skip Adair where they live. It is situated on the side of a hill about three miles from the field and overlooks the whole valley a splendidly situated spot. We had a bit of tea, then went to our hostel, where we were to stay. It consisted of about six wooden buildings of six rooms each, holding two to a room – not too spacious but comfortable. After supper we went back to their house where we met two members of the British Embassy; an American who we called Sak; and Dave Olsen. Sak is an American citizen though, born in Colorado, went to the same military school, New Mexico, as Greene, so these two had a few things in common. Sak also went to Colorado University and to Harvard. He almost has his Ph.D. He was about 35, but has the wit of G.B. Shaw. He traveled over quite a bit of Europe and Japan, and Far East. He is somewhat of a Socialist, I think, and is very much interested in China. Very interesting to talk to, especially Political Science. They brought several bottles of Scotch, which is at a premium up there, so a good time was had by all; except I wasn't too well so did not indulge. That night I also had the runners. I slept all next morning and went walking up in the hills in the afternoon, lying on the top, listening to the wind whistle thru the pine trees. That evening the gang of us got together again for a friendly game of Black Jack. No more liquor was had! Saturday, Sawyer and I spent the forenoon out tramping with our 22's, but did not get any birds. After noon the six of us, Sak and Olsen left to find the rest of their convoy they lost somewhere, went duck hunting. We tramped at least 10 miles and only got one goose and two teals. I got one of the teals with my trusty 22 after stalking him for 30 minutes on one of the lakes. That evening we were invited to Capt. Carney's house for a real Chinese dinner. His woman, Rose, a pretty Chinese woman, fixed up a most delicious Chinese dinner, and at first I had quite a time with my chop sticks, but soon caught on. Thereafter, I enjoyed it immensely. The main dish was chicken soup with egg foo young; almond dish, sweet and sour pork, chicken paddy; cabbage; two meat stews; and several other dishes. It really was good and I had my fill. The evening was spent playing records and drinking the Captain's whisky.

Sunday morning was spent mostly in bed again. After dinner we stopped at Loanes, Shepards, and Shemlin's house till we heard the Beechcraft coming in for us. After Hennesy and Mickelson, the pilots had dinner, we took off, and what a take off. We just cleared the trees and just tipped the tops. Not enough oomph in the old bus.

We got back at Kunming about 3 o'clock after a real restful sojourn ready to get back to work. I would like to spend about a month doing what we did up there – rest, go duck hunting, and a few delicious Chinese dinners. Nothing for us to do till night alert – Monday night.

Seems that the second squadron down at Rangoon base had a little combat. They had one raid which shot down fire, but Bright, Paxton and Christman were shot down, but not hurt. On the strafing job at Tak, Newkirk, Hill and Howard got seven ships – no damage to self. They went back with nine 40's and five Brewsters, got 24 at Tak, but Matt was seen shot down – not definite of outcome or him. Hope for the best.

The Japs now call use cruel, barbarous, blood thirsty Americans. One of the pilots shot down called us pugnacious, unorthodox. The radio at Tokyo also said the AVG is Japan's No. 1 Enemy.

Monday night Greene and I were on alert with nothing more to do than catch up on a little sleep. Having Tuesday off, several of us got a car and took a drive to the other side of the lake where a natural landslide left a sheer vertical drop of over a thousand feet. Many large boulders measuring 30 feet on a side were along the river's edge, some partially submerged. There was a small village amongst the rocks where the people lived that broke the rocks into smaller pieces, loaded them on sampans and sailed across the lake to the city for disposition. There also seemed to be some kind of kilns there that burned the rocks and gave forth voluminous black, sulfur odor smoke and left a black coal like residue. We could not find any interpreter around so could not find out what process was involved.

The next day, Wednesday, I was on day alert and ran several test hops, which I needed badly after so long a lay off. The next two days also passed along the same, but on Saturday, at about 10 o'clock, we were told an observation ship was flying near Mengzi, so we all took off, but all but four were told to return. McMillan, Shilling, Older and Haywood were the four who went to investigate. We were told to go to 20 angels over the field, but soon landed as we found out our four ships had intercepted three bombers. Shilling returned first stating he saw one go down for sure; soon after the rest returned and said two for sure and one probably. A little later the third was confirmed; so we hit 100% that time. That should vex the Japs to no end.

We heard over Tokyo radio that they had Mott in a hospital pretty badly burned up, but alright.

After the raid on Tak in Thailand, the 2nd Squadron down at Rangoon reported Moss missing, but he turned up several days later at Moulmein. Another piece of good luck.

The 1st Squadron finally got down to Hanoi for some pictures which turned out fairly well. They only had about 28 bombers and 15 fighters down there.

On Wednesday the 1st Squadron escorted 18 Chinese bombers (Russian SB-3) to presumably Hanoi, but because of overcast skies, went to Haiphong and bombed. They could not see their objective so could not get any results. All the fighters returned, but two bombers had some trouble and did not get back. The only opposition was some anti-aircraft fire.

A few alarms kept us in the air on several days, but a photo mission over Indo China, usually Hanoi, was the most flying any one put in. If it weren't for a test hop every other day or so, I'd be low on time and out of practice.

I had my first "Jing Bow" or Air Raid Alarm here at Kunming while on the ground away from the field. I had a day off and was writing a letter when I heard the siren and saw two red balls hoisted on the pole, which meant immediate danger. I took off with Tex Blaylock, our chief, who was at the dentist, into the hills about a mile from the hostel. As most of the hills are covered with mounds of grasses, it makes good bomb protection. They dig holes about four feet deep, large enough for one or two persons between the mounds and it affords the best protection next to a steel shelter. The only bad part of our hideout was that several anti-aircraft guns were located on top of the hills. As we were very near this hideout, we got situated and I could watch the multitudes streaming out of the city. The usual one way Burma road was packed with cars, trucks, rickshaws, carts and what not, taking up every inch of the road, and seemingly moving outward a few feet every little while, but surprisingly enough in fifteen minutes the road was clear along with the continuous blare of horns as everyone tried to out blow each other out of the way. The people on foot streamed past the traffic and many took refuge among our graveyard. There was room for plenty but plenty there were.

To our relief, the alarm was false and the only planes we heard and saw were our own P 40's out on patrol.

We were still hearing of the good work the boys down in Rangoon were doing. Twelve of Sandell's 1st Squadron went down for assistance, arriving a little late for one raid, but getting one the following day. It seems the Japs are only sending over Army 96's and 97's – no model "0"s. Bombers are single engine 98's only once did they send over seven 97's and they lost all of those. They are also coming over in small formations at night, but one of the R.A.F.'s Hurricanes got two of those. Then they are supposed to have about 10 Hurricanes down there and expecting more. From the many fields in Burma, the R.A.F. are bombing Bangkok with Blenheims and Lockheed "Hudsons."

Today, the 30th, I got back a letter I mailed to Mary Jane Steckmort on September 4th at Toungoo; it went to N.Y. arriving the 24th, and as no forwarding address was left, it came back here!!! Twenty days there and four months back.

On Saturday, January 31, 1942, the residing pilots of the AVG at Kunming were presented with their wings from the Chinese Air Force. A representative of the Generalissimo presented them to us as we held our first dress formation since our being here. Colonel Chennault expressed his thanks for us.

February 1942

Ever so often in the Chinese army, they hold a "comforting day" which consists of speech making and presentation of gifts to bolster the morale of the troops. We had one on Monday, February 2, at which the commanding officer of the Chinese Air Force and the president of the China Aviation Association, along with other members of the association presented Colonel Chennault with a scroll and embroidered silk American Flag. To 10 of the pilots present were given an embroidered silk bed spread for their good work under fire. Only those who had victories confirmed were presented, which left me out as somehow or other my claim was not substantiated.

We had a surprise on Wednesday afternoon when the CNAC] landed and the Generalissimo and Madame Chang Kai-Shek alighted. They were here only about half an hour; looked over the field, talked with Colonel Chennault and left for Lashio. We didn't get introduced to either, but were with them. The British Ambassador to China along with several other high ranking Chinese officials were also aboard.

The past few weeks went by without any excitement here except for several mornings when some of the better Chinese pilots were checking out in the P-40's. Of seven, only one had difficulty and ground loped, doing considerable damage to the plane.

The 2nd Squadron men were dribbling back by planes, the Beechcraft and CNAC. Squadron Leader Sandell was killed down in Rangoon while testing a ship. The exact way is not known. I hit the jackpot one day when I received out eight pieces of mail, six being Christmas cards.Sunday, the 15th, while starting my day off, we were told to report to the field. We really thought there was something up. Upon reaching the ready shack we were told that there were 50

Robert and Crew Chief Frank Losonsky
(Courtesy Virginia S. Davis)

P-40's for us at Cairo, Egypt and that we were to ferry them here. Six of the fellows were to go the following day. McMillan, Smith, Greene, Older, Haywood, and Laughlin were the first fortunate six. They left by CNAC for Calcutta on Monday morning. I was put in charge of operation along with Engineering so had my hands full. Monday was also the wedding of Petach and Miss Foster, a long awaited event.

Wednesday will long be remembered by those members of the 3rd Squadron who scored victories down at Rangoon, for on that afternoon they were presented with medals from the Chinese Government for their bravery and good work. As several of them were on the ferry trip to Cairo, only six actually were present to receive their reward. Olson was given a 5th class medal for his fine work on the ground, handling the squadron in a magnificent manner. Duke Hedman received a 5th along with another five-point star medal for scoring five victories. The others were Dupouy, Read, Jernstedt, and Overend. They received 6th class with a small pair of gold wings with the number of stars representing the number of victories. As neither Hodges' nor my confirmations came thru till Friday next, we did not receive any thing at that time.

The following Sunday another six pilots of the 3rd Squadron took off for Cairo or points west to ferry back ships. As the Generalissimo and the Madame landed here on Saturday night and expected to stay about a week, we had to keep the more experienced pilots here for protection, just in case. For that reason, the logical ones to go for the ships had to remain here; instead Shilling, Bishop, Foshee, Cavanah, Adkins, and Raines left.

The 2nd Squadron men were coming in from Burma now in greater strength so the pilots were put on the alert with us while their mechanics worked on their planes.

Things were getting pretty hot down in Rangoon, the Japs had forced their way across Moulmein, the Salween River, and now were right east of Pezu, which is only about 50 miles from Rangoon. Almost everyone had evacuated Rangoon, including the R.A.F. and left only about 12 AVG planes and three R.A.F. The rest went up to Magwee where nothing was prepared for the AVG and the British were evacuating that. Looks like the AVG will soon have to evacuate Burma as they seem to be left holding the bag.

On Monday last, the 23rd, Olson, Jernstedt, Hedman and Reed went out on a photo mission over Indo China and the two elements got separated somehow. Olson and Jernstedt got back O.K., but Hedman and Reed shot forced landings on a river bed some place in the vicinity. They did a good job, Hedman damaging only a prop, but both able to take off with a gas supply. They had radio contact, but we could not locate their exact position, although the Chinese were supposedly to know their whereabouts. So far today, being Wednesday morning, they have not been located by us or nothing done to get them back. Olsen tried looking for them from the air but did not succeed in seeing anything.

Thursday afternoon, shortly after five, fellows came up from Rangoon to trade planes, in came our two prodigal sons – Duke and Reed. They got here about 2 o'clock. They had a fairly decent time of being lost, only about 80 miles from here. We later found out Olson had flown directly over them but did not see them.

Saturday, February 28th, was a momentous occasion for the AVG The Generalissimo, Madam and several high ranking Chinese officials gave a dinner for us at our hostel. I was picked as aide to General Chow, the chief of the Air Force. All members of the AVG were present in the auditorium for supper and entertainment. The Generalissimo said a few words of great praise of the AVG as well as did the Madam. Both people won over all the command with their pleasant personality and vibrant energy. They are both forceful persons and have their hearts in their work. The Generalissimo's speech was translated whereas the Madame spoke in very good English. Then Colonel Chennault acknowledged their offerings and accepted their good wishes. Among those present were eight Chinese ladies who acted as hostesses. One rendered several fine songs in English and Chinese. Finally a Chinese play was put on which was humorous and very picturesque with many beautiful costumes. It concerned a fisherman and a racket with the poor innocent fisherman's daughter giving the racketeers hell – all parts were played by young boys. The squeaking voice of the feminine character struck everyone's humor.

March 1942

The following day six of us were to escort the General's planes to Lashio, but I had a rough motor so did not go and Overend's prop went out. The four escorted for about a hundred miles and returned just before dark.

Monday, the original six of us went to Loiwing by way of the Auxiliary airfields, Tusung, Yunnanyi, Pao-shan, etc. We got there about noon and were taken to the CAMCO quarters about 10 miles from the airfield. The altitude is only 2400 feet so it was very warm there. CAMCO has a wonderful set up with about four hangars for plane repair, about six large bungalows for occupation and a beautiful club house upon a hill. The club house is rather large with rooms on either wings, private bath in each room, three large bay windows 10'x15' overlooking the valley. It is all beaver board with teak wood floor and woodwork. The furniture is modern overstuffed and they have a wonderful 24 record radio-phonograph. The food was super and everything too good to last.

On Tuesday I had to fly back to Kunming to deliver a message to Colonel Chennault, taking 1:15 there and 1:30 back. Jernstedt relayed the answer from Loiwing to Lashio. That night the Colonel came to Lashio on the CNAC to see the Generalissimo and on Wednesday morning four of us flew to Lashio at 7 a.m. and escorted the General, Colonel, and party back to Kunming. Took two hours cruising at 140.

On Thursday morning, the General and party left for Chungking amid ceremony with a parade of six planes which escorted them up about 80 miles. Another six of us were to take off for Magwee in the morning, but because of the excitement did not get off.

Our little trip to Magwee was held up till Tuesday morning when five of us left at 11:30. Hedman was leading, with Fish, Reed, and Overend, and myself. We ran into hazy weather near Lashio and by some miscalculation, Hedman thought he overshot Lashio so did a 180 and started back. Went back for about 15 minutes then flew S.W. for another ten and then did weaving back North; by then we knew we were lost. We hit a valley with a river, railroad and road which we flew down for almost 10 minutes. Duke thought he recognized it as Bhamo so flew 140 degrees to hit Loiwing, but we were wrong and came back to the valley. This time Reed, Overend, and Fish flew about 60 degrees and Duke and I flew 140 degrees again. When we had about 15 gals left, we headed back toward the valley once more for a forced landing. We were told over the radio we were at F-3; S.E. of Loiwing, but I did not have a grid, or even enough gas, it did not help me any. When we got back to the valley with about 5 gals of gas, Duke buzzed several fields and finally picked out a sand bank along the river. He overshot the first pass and as I was so low on gas, I landed next. I hit short and was rolling along nicely when I hit a mound and bounced up a foot or so. The landing gear horn sounded, but the wheels tested O.K. so I pulled off to a side and watched Duke land O.K. We got Pao-shan on the radio, telling them we two were O.K. but know nothing of the others. When we landed there were only two natives, but soon had a large group of Burmese, one who was an interpreter and told us where we were at – 80 miles on a direct line to Lashio from Kunming – all we had to do was follow the river or road in the valley to Lashio – several Burma Frontier Force soldiers took guard of the plane and a Mr. Lee, Chinese, of the Vernon Burma R.R., who soon arrived, took us to his station along the road to Kunlong about three miles away. I was wearing poor fitting boots and got a large blister on either heel – tough walking! He served us some coffee, cookies and can of fruit, while we waited for a car from Kunlong. Two trucks of Chinese soldiers also came as they thought we might be Japs, but several were placed on guard over the planes. Duke and I had a six mile ride over a typical Chinese-Burma road to Kunlong, arriving about 5 o'clock; we landed at 2:45.

Kunlong is a construction camp along the 413R, its main station with Mr. Change, a Chief Engineer. There were about 250 men, several college men from the U.S., Civil Engineers. There were also several Americans who were Malaria control workers. The only one there was a Mr. Wright of the Rockefeller Foundation. Spent quite some time listening to his exploits during the last 35 years. He was in 32 different countries on malaria control. We stayed in Mr. Chang's building; which consisted of one central main room and four adjoining bedrooms. Duke and I had one together. We tried getting the 24-F at Lashio to get us gas, but did not succeed that night at all. After supper, a very good one, we heard that two other planes had landed up the river at Ming Tung about 20 miles. We did not know who or anything regarding the fifth plane. The evening was spent discussing world problems with Mr. Chang, Wright, Mr. Liu, the company auditor, who was in the 1919 Washington Peace Conference, and Mr. Mar, Mr. Wright's assistant. The company doctor fixed up my blisters.

Did not sleep very well that night, but had a good breakfast next morning at 7. About 10 o'clock a plane buzzed over, too high for us to see who it was, but he flew over our planes and the others. We continued trying to get gas from Lashio; but nothing but promises. A little later heard two more 40's from Kunming to Lashio.

After a most refreshing shower about 5 o'clock, in walked the three lost flyers; Fish had landed about 30 miles up the river, and to keep from going over a cliff into a river, nosed over but did not get hurt. Reed and Overend landed together. Overend hit a soft spot and went on his back too. Reed hit a dyke and washed out his landing gear. They were about 20 miles up. These two were picked up by Dr. Hall, an American with the U.S. Public Health Mission, working with Mr. Wright. Fish traveled by boat and horse all night and met the other two the next morning. Reed and Overend contacted Pao-shan and gave their details. Next morning the three took out guns, ammunition and radio form their planes. Except for the loss of our planes, we had a happy group together that evening.

The next morning two R.A.F. boys arrived with a YBR truck, with 16-20 gallon drums of gas. We all piled into cars and drove out to our planes and filled them up with gas. Coolies carried five drums on poles to the planes. Upon a further investigation, I saw I had sprung my landing gear and did not think it advisable to try to retract my wheels. Duke and I got off safely with our 2000 feet sandbank runway. It was very hazy and had a little difficulty finding Lashio, but landed O.K. My landing gear held up alright. We went to the CNAC hostel for sandwiches and received a note left by Olsen. Hedman was to go back by CNAC to Kunming, rest, get to Magwee by planes or car. I flew up to Loiwing in the afternoon and again made out O.K. with my landing gear down. Prescott came up with me (on instruments).

Dave Harris met us out at the field and took us to the clubhouse. I stayed in Dave's room for the night. Next morning I flew the plane to the runway next to the factory about 2500 feet long.

I found out that McGarry had flown over our planes to drop some food and messages and got lost; washing out another plane!!! Prescott took off for Lashio and Burgard and McGarry for Kunming.

Upon checking the plane more thoroughly we found that the bulkhead at landing gear section was cracked in two places so necessitated changing wings. This meant a lay off to the following Thursday. Friday was a day of rest.

Saturday, I took up a ship, the 100th assembled, for a check, but found out the prop was loose and caused vibration. Spent most of the day just loafing around. Saturday I checked the plane again and flew to Lashio, to check up on some supplies for the AVG The prop was changed so the plane flew O.K. Reed had flown Duke's ship to Magwee, Overend drove a truck there and Fish flew down Monday in the Beechcraft, because of a bad foot and stomach trouble came back to go back to Kunming. While at Lashio, I met Hennessey in the Beechcraft. He was lost two days on his way to Lashio from Kunming. They came up here on Monday and then went to Magwee. They returned Tuesday. Monday I went on a little excursion into the hills with a .22 hunting birds, but didn't hit anything. Didn't fly that day, but worked on an engine for the radio station, motor generator with Ciora.

Tuesday morning I ran the last check on the 100th, it was poor visibility so I did most of my flying buzzing everything in sight. The Beechcraft came in during the afternoon along with a convoy of the last gang from Rangoon, Fox, McClure, Overley, Harpold, and a couple others. Really had a gang here for dinner that evening.

Wednesday afternoon, Hennessey took off for Kunming and Mickelson took the P-40 there. Fish also was along. About 7 o'clock they all piled back! They had been out 3:10, getting almost to Yunnanyi, but because of bad weather, returned. They stayed overnight and left for Lashio, where Mickelson left the P-40, at about 2 o'clock Thursday afternoon. My ship was pushed out of the factory, but ran rough so could not fly. In the afternoon, DC-3 on CNAC landed with 18 passengers – among them General McGruether and General Naiden, several Colonels and General Maw. Quite a bunch of brass for one load. They all left Friday morning for Kunming. In the afternoon, I flew my ship from the factory to the field where they finished loading and hooking up the air speed. I did not have any [problem] taking off from the factory, but did not have any trouble. At about 3:30 I took for Lashio and Magwee, but when half way to Lashio my motor cut out and started running rough, I turned around and came back. We checked up and found out the gas was dirty, so they put new outside plugs in, cleaned the discharge nozzle. That afternoon Olson came up from Magwee. Next morning the weather was bad, so right after lunch we went to go back. He got off O.K., but my ship cut out on takeoff so I got about 200 feet and landed on the closest runway 60 degrees from the way I took off, making a 120 degree turn and landing, going over some railroad tracks, but not doing any damage. Olson got out ten miles and the weather closed on him, so he came back. After a short rain, it cleared a little so he took off again at 4 o'clock and got as far as Lashio where he stayed. I checked my plane again after they changed the inside plugs. It wasn't too good, but was alright.

Robert in China (Courtesy of the San Diego Air & Space Museum)

Sunday morning I got up early to get a good start, but the weather was really closed in. I did get away about 2 o'clock and took it easy as the engine did not run smoothly. It was rather hazy at Magwee so I could not get a view of the field till I was right over it and to my surprise I saw the field was in ruins. The planes on the field were just charred messes, some of them still burning. The hangars were half up, no personnel was on the field. It sure looked bleak and deserted. Quickly I scanned the skies for any sight of enemy aircraft or a fight. To my good fortune there was no action above me, so I buzzed the field several times checking up on bomb craters on the runway. I found a good spot and landed. Olson and

Reed were on the field checking up on the damage so they took me to their barracks in town. I found out they had a raid less than an hour ago and had no warning. About 60 bombers and 40 fighters came over and really cleaned up. There were no casualties that day, but on a raid the previous day, Johnny Fauth, a mechanic had his arm blown off and died. Swartz, Pilot, had his hand mangled, and Seiple, Mechanic, had a concussion; all three were caught running between trenches. That day, eight of our fellows got up, Jernstedt was hit in the windshield and got glass in his eye and face, but is O.K. Dupouy also had his windshield hit, but only got glass in his arm and shoulder. There was also an early morning raid on Sunday along with the afternoon, when nobody got off. The Japs really got us flat footed without any warning. I got gassed up and had a look around before I left. There were about 20 Blenheims which were all damaged beyond repair; most of them had bomb loads so they burned to a crisp. We had 12 P-40 of which about five burned, several were full of holes form strafing and four were able to fly up to Loiwing Monday. The Japs dropped mostly anti-personnel shrapnel bombs so did a lot of damage to planes. Luckily none of our personnel were hit. The RAF had already evacuated and our fellows were leading up. They left early Monday morning. That spelling the end of Magwee and probably Burma. I made out O.K. on my return trip to Loiwing, landing here about 6:40, just before dusk set in. A group of eight planes from Kunming were here, they were on their way down on strafing missions, there were two more at Lashio. Monday morning, four ships flew up from Magwee, Hodges, Prescott, Mass and Wolf. We sure were surprised to see four ships come up from that mess. In the afternoon, Neal and his gang took off for their strafing job. They landed at a field near Heho in Burma, refueled and left early the next morning for Sinkiang, Indo China. They split into two groups, Neal had six and Newkirk four. Neal hit the airfield at dawn and caught about 40 fighters lined up just perfect for a good strafing job. With about four passes, they got everyone. Later Newkirk came over but nothing was left for him so they strafed some troops and armored cars. One of the cars got Newkirk and he went right in. Another squadron leader lost to the AVG. Black Mac McGarry was hit by anti-aircraft fire, but flew for about 60 miles before his plane caught fire and he bailed out in no man's land between the two lines and about 30 miles east of the Sittang River. The other fellows tossed him a map and he took off for the woods.

That morning I also went to Pao-shan, 130 miles N.E. of Loiwing, to get a radio transmitter instruction book, so our boys could get our station going. I left without any breakfast and as soon as I got back, there was a Jing Bao, and as Neal's flight had returned and had not much ammunition left, we all took off to clear the field of planes. I went to Pao-shan again and stayed for several hours and returned here about 5 o'clock. That evening Olson and Dupouy came in with a car from Magwee, and started the incoming of the rest of the convoy which staggered in.

Wednesday we had a four ship alert here, Neal, Bond, Prescott, and myself, but nothing happened. Hodges returned from Kunming where he had gone Tuesday from the Jing Bao. In the afternoon, Neal left with eight others for Kunming, leaving us one good plane and four in repair.

Thursday, Olson went to Kunming and returned with seven other ships. Most of the mechanics came in and got settled in their barracks near the field. We moved into one of the bungalows below the club house and I lost my restful homestead "upstairs."

Friday the 27th, we had an eight ship alert and had one alarm at noon. But sighted nothing and only flew 45 minutes. The next few days we had patrols and alarms but except for one morning when Older and Greene ran into a lone observation ship, and Older got him on his first pass.

Wednesday, April 1st, Shilling brought a P-40E here, so we had our first look at our new ships. We thought they were pretty nice.

Nothing much happened the next week, except a few false alarms and one day Lashio was bombed twice, but we received the alarm too late to do anything. The following day Mandalay was bombed.

April 1942

The weather started closing in and we had several days of light rain. Sunday, April 5th, was Easter Sunday, so we had Easter Church Services in the club presided by a R.A.F. Chaplain. We were fortunate in having some moving pictures left here enroute to Kunming, and we saw the first Monday night, "Here Comes the Navy."

Also received about six Christmas cards, a letter and a telegram from Skid. More 40 E's have been coming thru and met a classmate of mine now on the PAA ferrying the ships.

Thursday was my day off, but spent the morning out in the hills due to an alarm, and then in the afternoon when we had another alarm, something told me it was the real thing. About 12 Model "O" fighters came over on a strafing job, but made only two passes on three ships which were left on the field, a Blenheim and two 40E's. Then eight of twelve ships we had up, they were all up over 25,000 feet, jumped the fighters and accounted for ten Japs, several of them right off the runways. None of our boys got hit but two hurry birds ran out of gas and had a forced landing. It was the first fight over Loiwing in several years and we were the victors. Seven 40E's and three 40's came in (Fri.) The next morning about 6 o'clock several of us still sleeping in were awakened by the sound of engines and within a few minutes hear the slow chattering of machine gun fire. From our room we could look toward the field and saw five fighters making passes at our field. Our alert pilots were half way to the field and several of the mechanics were in their ships, warming them up with five Japs came *(sic)* over on a dawn strafing job. The mechanics got out of the ships and hid safely so there were no casualties, but of 20 planes dispersed over the field, the Japs only damaged seven badly, but not beyond repair and about four which were only barely hit. None of our planes burned. As we did not have any planes in the air, the Japs had a circus and no opposition, but still did not score too great a victory. We carried out two reconnaissance missions over Toungoo area and also had another morning alarm. That afternoon at 2:45 we received an alarm and took off. Five of us got together at 25,000 feet and were cruising for almost a half hour when we were told there were planes over the field. We started down, the first two planes diving below a layer of clouds, then the second two, of which I was No. 2, saw planes above the base of the clouds. I pulled up sharply and got my sights on a lone Jap in a head on deflection shot. My sight was too dim so I used my tracers for aim, but did not get him in that pass. We then tangled up a bit, I getting in about three good bursts from all angles. On the last pass, we started head on. I opened fire and he turned to my left. I followed him still firing and then saw him turn over on his back, flames shot out from under his wing and he started down. I followed him a short distance and saw another plane to my right. It was another 40 so I pulled up and went into some clouds to clear myself. I flew over and under the clouds by could not see another plane in the sky. I climbed for more altitude and cruised mostly over the field. About five minutes later I saw a plane going south over the mountains and took off after him, but then saw another one below and behind which was closer so I turned after him. After a five minute chase I got close to him. He dived steeply, I followed, but as he started pulling out, I saw the Chinese star so it proved a wild goose chase. By then the scrap was over, so I landed, the third back, hardly got out of my plane when we got another alarm, but we took off into the hills and the alarm proved false. We counted the score, got four positive, three probables; the R.A.F. got two, but lost two ships; one pilot forced landed O.K. The other bailed out. Not one of our planes got a bullet in it.

Older and Hedman had a 100 mile dog fight with a Jap who proved to be a real pilot, but they finally bagged him.

Saturday morning we had *(sic)* up at dawn patrol, but nothing came over. It was my day off and had to spend the morning in the hills on a false alarm. That afternoon I wrote four letters! Hope they bring in returns. The 2nd Squadron which came down here from Kunming sent out two patrols of three ships each, 40E's over Toungoo area, but did not run into anything. We had a movie that night, "The Mad Empress."

Sunday, the 12th, I was spare pilot so got out a little late for the dawn patrol, but again nothing came over. Hill, Wright, and Croft were down over Toungoo and Tex Hill and Wright staffed the airport. Tex got three and Wright shot down an observation ship.

Nothing else happened except perhaps that I checked out in a 40E that afternoon. I think it is a pretty good plane, but did not get a chance to put it thru its paces.

The next few days were filled by alarms, but no enemy ships ever came over. Some days we would have four alarms and never make any contacts. However, on Saturday, April 18th, the third alarm at about 11 o'clock proved to be rather interesting as far as I was concerned. We had a flight of six, led by Dupouy. I had the second element with Prescott on my wing and Haywood the 3rd element. At about 18,000 flying west over the field, I saw a twin engine, light colored plane with red circles on his wings. I tried calling Parker and then flew up close and wobbled my wings. He did not see me and the enemy plane was pulling away at about 60 degree angle to the north, 4,000 feet below, so I did a sharp wing over and started down. I dove behind and below, coming up at about 300 yards and started firing. My tracers showed I was firing low, so I brought the fire up till I saw it was going into the ship – the right engine started pouring smoke and I had to dive away to keep from running into him. By the time I got on him again, Prescott was on his tail, followed by Haywood. The ship was leaving a long trail of black smoke and diving down at over 350 m.p.h. about five miles away he finally did a sharp turn to his right, did a two turn spin and then dove into the ground with the right engine in flames. When he hit, a very large ball of flame rose 20 feet in the air and died out. I buzzed the spot but there was no wreckage left – it was all over the place – no part very large. We got together again and stayed up for another half hour before landing. Prescott and myself took credit for the one Jap observation ship (1/2 each!)

That afternoon, we discussed several missions we were asked to do and thought them too risky so started action against them. We had a meeting that night with Colonel Chennault, who was made a Brigadier General in the U.S. Air Corps. He gave us no satisfaction, so we signed a letter of resignation. 28 of us signed it. The next night, we had another meeting at which he told us he would not accept our resignations, but consider us as deserters if we left. He appealed to our patriotism, courage and everything else he could think of, but the desertion part seemed to change most of the fellows' minds. We decided to stick together and follow orders and see the thing out.

Monday, the 21st, Dupouy, Groh, Laughlin, and myself went down to patrol over Pyknic in the afternoon. All we did was dodge anti-aircraft fire which gave off black puffs of smoke about two feet in diameter and they would come pretty close too. Everything seemed hundred feet at the same level. *(sic)* We would twist and turn, dive and soar, never flying the same pattern for more than four or five seconds. We did not run into anything nor see anything on the ground, so we went back to Namsham to refuel and stay overnight for a return patrol the next morning. We landed at about four and at 4:30 saw an observation ship 10,000 feet above the field looking things over. As soon as he left, Laughlin and I took off just in case anything would come back to strafe. My oil temperature hit the peg and dropped down, so I had to land. Link stayed up. About 6:30 Older and Shield came in to pick off an observation ship that was in the habit of coming over in the morning.

The British had evacuated the place so there was nobody there but our own little gang and luckily two cooks who got something together for us.

Olson and Rogers worked on the planes all night and got them in shape for the next day.

We got up before dawn, had somewhat of a breakfast, got the planes warmed up. The canopy on my plane had blown off and could not be repaired so I was not going to go on the mission, but go back to Loiwing instead.

The three fellows took off at about 6:30 and got on their way. We pushed Shield's and Older's ships into the hangar and I went to take off.

The takeoff was normal, but the oil temperature started creeping up so I circled the field a few times and when the temperature hit the peg and the pressure dropped, I decided to come in. I landed way from the hangars, pulled off to a side to park the plane, but then thought it would be better if I put it in the hangar for the fellows to work on, so I started racing back down the runway. As I approached the taxi strip to the hangar, I started to turn when I saw white flashes and hear noises coming from the right side of my engine. My first thought was that my engine had caught fire.

(This should be the end of the diary, because by all rhyme and reasons, I should be dead and not writing any more – but will continue in the next book.)

--- PART TWO ---

After the flashes in the engines, they moved into the cockpit – I thought my guns were going off in the cockpit. The flashes blinded me, I felt sudden pangs in my legs, I heard a roar, I looked up and saw a plane sway low over my head, I knew I had been strafed. With the plane still moving very slowly I coolly unbuckled my safety belt, climbed out of the cockpit, felt my head set plug in jerk out of the socket, jumped on the ground, out of the way of the oncoming tail, and started hastily to take off my chute. The first three buckles came open easily, but the fourth stuck, as I played with it, I saw several things. First a small hole clean thru my thumb below the first joint, second a plane making a split S to come down and strafe and another formation of four planes flying overhead.

Luckily in not being killed and in getting out of the plane and the second in getting out about 20 feet from a covered dugout, which after I threw my chute off, I dove in head first, along with two Chinese soldiers. It was not till then that fright or pain overtook me. I could see blood coming out of my left shoe, so I took it off and saw a hole only through the top part of my sock. I took that off and saw a ragged wound. I could see the missile and tried forcing it out, but soon gave up. By then I felt other pains in my legs so pulled up my trousers and saw about four gun shots in each leg, all seemed to have stopped inside. I tore up several handkerchiefs I had and with the help of one Chinese soldier, tried making some sort of bandage to stop the bleeding.

All this time a continuous drum of machine gun bullets were splitting the air. My plane was on fire, the gas tanks giving off a dull roar, then gulping pants as it burned. The oxygen bottles gave way and sounded like a real high pressure force going off. Then the machine gun bullets started going off, first rapidly then intermittently, and soon the enemy planes left and only the flaming plane and shooting bullets broke the stillness of the morning. I tried to send one of the soldiers for the other fellows, but they came shortly. They were very happy to see me and though me a goner.

They put me in the back of a truck on a stretcher and took me to the hospital. Since the British had left there were no doctors, so Cross went to Loiwing and was very fortunate in getting Dr. Seagraves. We did not know this and Shield had started to drive me down there but we all came back. The doctor had two native nurses and at exactly 12 o'clock I went under chloroform. I came out at 3 o'clock, really cussing at the Japs. The doctor had left. Shield and Older left for Loiwing and Cross received orders to drive me to Lashio to be picked up by a plane.

I had a cast on my thumb and palm, bandage on my left ankle, calf and knee, and two on my right leg. I did not feel very bad. We left at 6 p.m. and drove till 10 when we got a flat and had to wait till morning to borrow another tire. We had a fairly rough ride of 240 miles to Lashio, getting there at almost 3 o'clock. They took me to the field, but the plane was late so they took me to the R.A.F. hospital in Lashio. There I met Dr. Klein from Toungoo who was a British officer now on his way up.

That night was a night of torture on a hard wooden bed, not sleeping or pain-killing powders. *(sic)*

The next morning they took some x-rays. We left by Beechcraft at about 1 o'clock for Loiwing with Dr. Richards who had come down to see me. I spent the night in one of the bungalows with Engle who was getting over the typhoid. Next afternoon doc took me to the hospital at Loiwing to take out some of the shrapnel they found with the x-ray and fluoroscope. They tried spinal anesthesia, but it wouldn't work, so they used ether. They used it twice as the fluoroscope after the first and found some more pieces. I came out at 8 o'clock and spent another torturous night on a hard wooden bed with no pain relief or sleeping pills. They picked out five pieces of shrapnel from the two legs, one from my right knee joint which might have proved fatal in the future.

Late that afternoon, they took me to the field, put aboard a DC-3 (CNAC) flew to Lashio and then to Kunming – got here about 9 o'clock and taken to a hospital where so far everything has been pretty good.

Dr. Sam Prino took charge and changed the dressing and also put a new cast on my hand.

The Jap mechanized spearhead took Namsham and went up and took Lashio. On several fighter sweeps the fellow ran into observation ships below Lashio. One day they did several strafing jobs on truck convoys. On one occasion, they ran into a formation of bombers, escorted by fighters going to Loiwing. They got about 12 fighters but the bombers went on and bombed the field at Loiwing.

The squadron moved to Mingshik and started evacuating Loiwing for Kunming as the Jap army was moving up. In several days everyone evacuated both Loiwing and Mangshi.

May 1942

On May 4th, the first squadron was at Pao-shan when bombers in two waves came over and bombed the city, killing Ben Foshee. Charley Bond took off, chased the bombers, got one, but when he came in to land, was jumped by fighters and his ship caught fire. He bailed out, but was burned on his face, shoulders and hands. He came to the hospital the next day. The Japs also came over Pao-shan on the 5th, but we had seven planes up and got about nine planes.

The next day we found out Peterson was turning back and could not take us on to Karachi, so we would have to wait for another ride.

That afternoon we looked up "Pappy" Greenelaw and we went to a jeweler in Old Delhi where I bought a diamond and emerald ring and bracelet to match. A really beautiful set! That afternoon, I went swimming at Hotel Cecil, where Pappy was staying with Edgar Snow. Had a fine time, but almost overdid my exercise. McMillan got in that day so they went shopping and I met them at the hotel that afternoon. Spent the evening at Pappy's with Jernstedt who took Pappy into town with Acey-Deucy. Met "Mouse" Moore at the airport where he was control officer for that district. Another group of fellows or rather Jo Steward and McHenry got in the next day, spent the afternoon and evening with Joe and Mac.

We finally got off with Nowak as a pilot on a none-too-good ship. It was overloaded so could not go very high and had a rough and long slow trip. We got into Karochi about an hour late; in fact a ship that took off after us beat us in.

At Karachi we met some of the fellows who left Kunming before us and learned that some had gone on to Bombay to catch a boat home. There seemed very little chance of getting a plane out of here in the near future. I met Red Hall from Mitchel Field, so lived in his room at a cottage on the post. Belden was also stuck here awaiting a plane to Dingin.

Frank Schiel was missing, but turned up several days later at Yunnanyi. He had run out of gas and had a forced landing.

The Japs were now at the Salween River, 40 miles from Pao-shan and the fellows went down for several days on dive bombing and strafing missions routing the Japs into a retreat.

Crew Chief McAllister ran into trouble and was caught on the other side of the Salween. He took a roundabout course and after walking two days, got into Pao-shan and finally Kunming. Hastey also ran out of gas and had a forced landing, but finally hit Yunnanyi O.K.

Wednesday the 13th, Jones, Shield, Laughlin, Donovan, and Bishop, went down to Hanoi in E's and made a one pass bombing-strafing job. They got 15 for certain and probably 15 more. Donovan was shot and crashed near the field. The other fellows were raised a rank by the Generalissimo, Shield and Jones were made into Squadron Leaders.

The following day fate played a dirty trick. Jones was on the bombing range practicing dive bombing and dove too low and stalled on the pullout. He spun in from about 1500 feet. A rather unwanted ending for Jones.

The next few days several flights went down and strafed the Hanoi-Lake R.R. and got two trains – again fate came thru – Bishop was hit by AA over Lashio and when his ship caught fire, he bailed out. He was captured by the French, but was turned over to the

Japs. They tried to ransom him from the French, but they would not cooperate.

Bob Little was the next to go. They were down on the Salween bombing artillery and bob got hit in the left wing by anti-aircraft fire. It just disintegrated as if his own bombs went off. He was pretty low so just spun in and burst into flames – a rather inglorious finish.

Made my debut in town by going to the movies one night and out to the field the following day. Sunday morning we had an alert due to Yunnanyi being bombed, but they did not come this way.

Sunday, May 31, the WASC gave us a lawn party in front of the Hostel. We had some delicious Chinese food and drank rice wine. Some of the fellows had a good amount so a good time was had by all.

June - July 1942

Monday I had my cast taken off and saw my crooked thumb. Was swollen a bit and could not move it hardly at all, but it looks O.K. May have a scar on each side where it went thru.

Not much happened for a while, had bad weather most of the time. The 1st and 2nd Squadrons could not go up to Chungking because of the weather.

They finally got me to go to work as operations officer on June 6th, and the next day the squadrons got away for Chungking. Also the following day we had reports of Japs coming up, but because of bad weather did not get here.

Wednesday night, June 10, about 10:30, a pilot of one of the DC-3's, Groh, along with Peret, the group engineering officer, came into the operations shack and started plotting a course to Hanoi. I tried my best to persuade them to call it off or postpone it, but they had their minds set on it. They took off at 12:20 with 800 gallons of gas and 3000 pounds of bombs. We traced their course to the border by the net but could not tell if they hit Hanoi or not. They said they dropped all their stuff on a city, but I think it was Haiphong or some other city, We never did get any confirmation from any news report. They got lost on their way back and hit way northeast of Kunming, but luckily enough we brought them in on the R.D.F. at about 7:30 next morning. I sure sweated them out.

Thursday afternoon I had a fatty tumor on my neck removed by Dr. Monyet, a Public Health doctor working with the group till the army comes in. He had quite a job removing it. I had a local so knew what was going on. It caused a little pain that night and today, but seems to be coming along O.K. Also heard today that the 1st Squadron bagged eight Japs near Kinelin.

Rested several days at the 1st Hostel before moving to the 2nd, when I took over Operations. Dr. Monyet took the stitches out about six days after the operation, so I was finished with that business.

Rains came and sort of held up my operations and the place got rather muddy.

Came the 27th, and the first detachment took off for home. I found out that I could get sick leave and go home earlier and perhaps avoid the rush, so Farrell and I went to the Mozir Gentry that evening. He told us to see him and the General the following morning, which we did and got an O.K. to leave. We did get away on a DC-3 that afternoon of the 28th at 2 p.m. The pilot was Peterson, a former T.W. boy, and did a very good job on instruments a good part of the way. We were up at 16,500 and I got a headache from lack of oxygen. We hit Dinjan with more rain about 5 o'clock. It was terribly muddy, but we were taken to the Officer's Barracks 15 miles away. The field looked very much like Toungoo except this had two finished runways. We stayed overnight in the upper floor of a tea drying warehouse on a plantation. The main building was full. I met Ham, a fellow student at Morton Junior College and class of '42 Flying School. Didn't think much of Dinjan at all and was glad to leave the next morning at 6 for Delhi. The first hour out we flew over the tree tops because of low ceiling, but finally made out O.K. I put in an hour as co-pilot. We arrived at Delhi about 1:30 at a very beautiful airport in a lovely city.

Delhi looks like a very well planned city before being built. Well laid out, plenty of trees and shrubbery and modern style buildings. The main attraction was the Government buildings and Viceroy's Palace. It was all of red brick, well styled architecture. Fountains and walks, shrubbery and trees made it a very pretty site.

The center of the circular is the Government buildings. On one side several miles is the airport and on the opposite side is the business circle. The center is a small park and the shops from*(sic)* around a circular drive around the park. The streets branch out from the park into residential sections. The homes are almost all yellow plaster finish, square, California style. A very beautiful sight form the air.

We stayed at the marina Hotel in town, having a double room with two beds in each. The hotel was used as the officer's quarters. Was very nice, including good food. We spent the first day looking the city over and did a little shopping. Had my first malted milk in a year that afternoon. In the evening we went to a pretty good theater. So spent some time sightseeing and going to show with him.

The city is very much like Rangoon, perhaps not as many newer buildings, but not so much of a native market place and a little cleaner and neater. They have several nice theaters of which I visited one.

Saturday, July 4, I celebrated at the officer's club in town, where I met Dave Wallace, old Pursuit Instructor from Langley Field and also Ray Boggs' old buddy from flying school, sister in-law who is one of 90 American nurses here. Windy Miller from Mitchel had her at the dance. Had a good time that night and slept with Wallace and Miller at their barracks, 51st Pursuit Group.

----- END OF DIARY -----

"What's Next" War Diary of Robert R. Brouk Notes

[18] Schultz, Duane. *The Maverick War*, p. 110.

[19] Bond, Charles, and Terry Anderson. *A Flying Tiger's Diary*, p. 39.

[20] Chennault, Anna. *Chennault and the Flying Tigers*, pg. 98.

Coming Home
July 1942 – November 1942

In July 1942, the American Volunteer Group disbanded and the pilots and crew returned to the U.S. to rejoin the U.S. Army Air Corps and fight in World War II.

Robert flew home from his service in China by way of India, Africa and South America. After arriving in Milwaukee, Wisconsin, Robert boarded the train to Chicago and from the Chicago train station, made his way home via the "L".

Upon his arrival home the local newspapers began running almost daily accounts of his comings and goings and life overseas. After an interview with the *Chicago Daily Tribune* it was written that "Robert recounts his experiences in a manner that is refreshingly modest and yet confident and alert. He is credited with knocking four Japanese planes out of the sky."[21]

The Berwyn Life reported, "Bob displayed the modesty that marked his whole career ......strolling through the old familiar backyard of the Brouk residence, and surprising his mother – who was busy preparing a meal in the kitchen – with a quiet, 'Hello Mom.'"[22]

Patriotic spirit was high and Robert became Cicero's "hometown hero" because of his engagement with the Japanese in the skies over China.

The Berwyn Life newspaper began planning a parade and ceremony to honor Robert which would be held on August 2, 1942. The newspaper ran articles almost daily in the weeks before "Bob Brouk Day," praising the heroic efforts of Robert, the Flying Tigers, and other local men serving in the Armed Forces in the war.

The July 22 issue of *The Berwyn Life* stated, "As a member of the Flying Tigers, he received a rating of Flight Leader, a citation from General P.T. Mow, head of the Chinese Air Force, and a silk scarf from Madame Chiang Kai-shek for his bravery under fire."[23]

Robert quickly began making the rounds to various organizations, events, and radio programs. In late July, the Air Force held Cadet Rallies in Chicago to inform and inspire men to join the U.S. Air Force. Air Force sponsors were instructed to recruit 20,000 men from the Chicago area before the end of 1942. Robert attended one of these rallies with fellow Flying Tiger, Edwin Fobes. The men were honored at a luncheon at the Merchant's and Manufacturer's Club before they viewed a spectacular air show over the city. Chicago's Mayor Kelly said of Robert and Edwin, "These men have the key to the city and we mean that. They have felt enemy fire. They are not thinking of their lives, but of the country....."[24] Later that week Robert spoke on a WMAQ radio program about the day's events.[25]

The Berwyn Life's July 24 issue reported on the progress of the "Bob Brouk Day" plans. The Western Electric Hawthorne Works Plant approved the use of their Albright Memorial Field for the celebration ceremony. This was important in the history of the Western Electric Hawthorne Works Plant and Cicero because it was the first time in seven years that approval for a non-Western Electric event was given. The company reminded the planning committee "that the field is dedicated to the 61 employees who gave their lives for their country during the First World War."[26] With this in mind, Western Electric Hawthorne Works felt using their field for the ceremony was in keeping with the patriotic spirit of the time in the country.

On July 27, Robert attended a reception in Chinatown with fellow Flying Tigers, Chaplain Paul Frillman and Sergeant Major Edwin Fobes. The *Chicago Herald-American* reported "Four thousand Chicago Chinese residents will give a tumultuous welcome to three of the famous "Flying Tigers....."[27] After the reception he, along with Frillman and Fobes, were interviewed on the radio station WLS.[28]

Robert's radio interviews continued on July 30, when he spent an hour being interviewed on station WBBM during the *Victory Matinee*.[29] His mother, Emily Brouk, was interviewed on August 6, on station WCFL during the *Our American Service Stars* hour.[30]

As the interviews continued to keep Robert in the spotlight, the "Bob Brouk Day" planning committee continued their plans for the celebration. They decided Robert and his parents would travel by car at the head of the parade route as military and civic leaders followed. No floats were allowed in the parade, however all Cicero clubs and organizations were invited to participate in the parade and ceremony. Some of the groups that participated included: Boy Scouts, a group Robert was very involved in as a teen; Girl Scouts; Morton High School Band; Local American Legion Posts; Bohemian Sokols; Cicero Fire and Police Departments; Chinese Boy Scout Drum and Bugle Corps; Cicero Post American Legion Drum and Bugle Corps; and other musical and civic groups.

Amid all this planning, Robert continued to visit various local clubs, businesses, and organizations. He spoke about his experience as a Flying Tiger. It was on one of these visits, to the Western Electric Hawthorne Works Plant that Robert met his future wife, Virginia (Ginny) Scharer. It is interesting to note that when Robert was stationed in Burma with the Flying Tigers, he purchased a Chinese robe for the girl he would marry. Little did he know that within a few months of that purchase, he would meet that girl.

Ginny, an invoice typist, was the 1941 "Hello Charley" girl for Western Electric. "Hello Charley" became a tradition after a post card addressed to "Charley, Western Electric Hawthorne Works, Chicago, Ill.," was sent to the plant. It was intended for a man named Charles Drucker, but the sender could not remember Charles's last name. As the postcard circulated the plant searching for the proper owner, plant workers started calling each other "Charley Western" which became "Hello Charley" as a greeting.

Hello Charley Sticker 1941 (Courtesy Virginia S. Davis)

Each year after the tradition began hundreds of women were nominated to be the year's "Hello Charley" girl. Elections were held in May and the "Hello Charley" girl and her court were crowned in June. Of course, promotional items followed -- including auto and luggage tags with the "Hello Charley" logo and the current "Hello Charley" Girl's photo, thus identifying Western Electric workers all over the world.[31]

Robert and Ginny met at the homecoming ceremony held by Western Electric Hawthorne Works Plant July 30, 1942. The event took place in the outdoor recreation area of the Plant where the employees caught a glimpse of the heroic ace. Ginny and another woman were asked to show him around the Plant and take him to lunch and make him feel at home.

Ginny described her first impressions of Robert. "He was so unassuming, kind and thoughtful. Therefore, I was flattered and excited when he said, 'Would you like to join me for a cup of chi after all this hullabaloo is over?' I joined him – we talked, then dated and you know the rest."[32]

As the month of July drew to a close, Robert spoke at a luncheon at Klas Restaurant to a group of about 80 Cicero Lions, Kiwanians and guests of service groups about his experiences in China. *The Berwyn Life* described his apparel as "neatly clad khaki shorts and a pair of shoes," a uniform worn in China. *The Berwyn Life* stated, "He described the Jap pilots as "methodical," stating that they lack initiative and cannot shoot straight." Robert went on to talk about the December 25, 1941, battle against the Japanese where the Flying Tigers fought against roughly 80 Jap bombers and 40 fighter planes while shooting down 26 of them without losing a man.[33]

The month of July quickly ended, and "Bob Brouk Day" arrived on August 2. Thousands of local citizens lined the parade route in anticipation. The parade started at 2:00 p.m. and proceeded east on Cermak Road to the Western Electric Albright Memorial Field at Cermak Road and 50th Avenue. The parade line up was as follows: Division 1 – Military; Division II – Veterans; Division III – Sea Scouts, Boys Scouts, Girl Scouts; Division IV – Sign Painters Local Union No. 830; Division V – Civic; Division VI – Fraternal and Social.[34]

Robert in uniform with his father Peter to his right seated next to his brother Harold before the parade. (Photo courtesy Virginia S. Davis.)

Hello "Ginny —
Best of Everything
Bob Brouk
"Flying Tigers"

At 3:00 p.m., as initial vehicles in the parade lineup reached the Albright Memorial Field, a ceremony opened with the "Star Spangled Banner" sung by a schoolmate of Robert's, Keith Smejkal. The anthem was followed by a prayer given by Bob's former chaplain in China, Rev. Paul Frillman. The prayer was followed by a speech by Mr. David Levinger, the Vice-President of Western Electric Hawthorne Works. Major Lloyd M. Showalter of the U.S. Army Air Corps followed and finally after many other speakers, Robert was introduced and interviewed. After Robert's speech, his parents were introduced before Robert presented Gold Stars to the families of Cicero's three known World War II Service fatalities. Following the presentation, a "parade in the sky" was held. Planes were flown from the Army Air Corps units, the Illinois Air Militia and the Civil Aeronautics Air Patrol.[35]

After "Bob Brouk Day," *The Berwyn Life* continued to publish stories about Robert as he was honored by many local organizations. He attended a Lions Club luncheon in his honor on August 7 at Klas Restaurant. Before Robert left for China in 1941, the Lions gave him a luncheon at which *The Berwyn Life* reported the Lions "filled him with a lot of raw meat, to prepare him for the fighting he was to

Robert, his father on his left, his mother to his right, his brother Harold in the front seat, (Both photos courtesy Virginia S. Davis)

do."[36] The Lions Club agreed the raw meat helped him fight the Japanese, so another platter was to be served to prepare Robert to re-enter the United States Army Air Corps and fight in World War II.

Robert participated in the Czech-American National Alliance fourth-annual Mobilization for Freedom parade on August 9. Participants watched a parade beginning at 26th Street and Karlov Avenue, and ending at 26th Street and Albany Avenue, before congregating in Pilsen Park for a War Bond Campaign. The gathering included not only Czech-Americans, but also local civic groups; federal, state, and local government officials; World War I Legionnaires; Sokol groups; and other patriotic organizations. Robert directed the war bond drive from the cockpit of an airplane placed in Pilsen Park for the event. He was assisted by women in Sokol uniforms for the duration of the afternoon.[37]

Robert at the Hawthorne Works Plant (Courtesy Hawthorne Works Museum)

Continuing his round of speaking engagements, Robert was the guest speaker at the Cardinal Council Knights of Columbus Meeting in Cicero on August 11. *The Berwyn Life* reported "Brouk's visit to the council is in some respects a matter of reciprocation"[38] due to the fact a Knights of Columbus group participated in the "Bob Brouk Day" parade and festivities. Then on August 21, Robert was initiated into the Cicero Moose Lodge in what became known as the "Bob Brouk" initiation class.[39]

The September 1942 issue of the Western Electric Hawthorne Works internal newsletter, *The Microphone*, featured an article about Robert and another Hawthorne Boy, "Chuck" Laver. The article discussed not only Robert's appearance at a noon-hour program at the end of July at the Hawthorne Works Plant, but also "Bob Brouk Day" which was held August 2 at the Albright Memorial Field. It was written, "The crowd of Hawthorneans cheered the young pilot enthusiastically when he appeared here and told of several of his clashes with Jap planes."[40]

Robert continued making the rounds visiting organizations and appeared on radio programs into October. He participated in a radio program called, "Wings of the Army" on October 9 on WGN Radio. Robert was one of many Army and Air Force speakers to aid in the war's recruiting effort.[41]

Robert rejoined the Army Air Corps and *The Berwyn Life* reported on October 12, 1942, that Robert was on his way to Washington, D.C., "in anticipation of orders which probably will transfer him to Orlando, Fla., as a flying instructor."[42]

The report was correct, and in October he was stationed in Orlando, Florida, at the newly formed Army Air Forces School of Applied Tactics (AAFSAT). Robert was promoted to the rank of Captain[43] and assigned to the 81st Pursuit Squadron, 50th Fighter Group, 10th Fighter Squadron[44]. Charles Bond, a fellow Flying Tiger from the 1st Squadron, served as the 81st Pursuit Squadron's Commander during that time.

One purpose of the AAFSAT was to train newly formed cadres in tactical combat flying techniques under simulated conditions. The cadres flew in and out of airfields with little infrastructure to prepare them for combat. Another purpose was to test and evaluate new tactics which could be used in any World War II combat theatre.

Robert was made a Flight Instructor due to his training and combat experience with the Flying Tigers, shooting down Japanese planes and being credited with 3.5 "kills."[45] Robert soon began training pilots for the rigors of aerial combat.

While Robert went to Orlando, Ginny remained in Chicago. The two fell in love quickly at the end of July when they met, a whirlwind romance. "As I look back now, it almost seems like God said, 'Hurry up, do things fast. You only have a short time left on earth.'"[46]

Robert proposed to Ginny on October 30, 1942, over the phone. He sent his parents a telegram announcing his proposal. Ginny spoke to her father about the proposal and he said, "Do you love him?" and after she answered yes, he said "All right, but remember it will be a permanent life style change."[47]

(Courtesy Virginia S. Davis)

With Robert in Orlando on duty until shortly before their wedding, it was up to Ginny to obtain the marriage license. *The Daily Times, Chicago*, reported on November 25, "She went down to the city hall yesterday and obtained a marriage license for herself and her fiancé, Capt. Robert Brouk..."[48] The two were to be married three days later on November 28, 1942 at 8:00 p.m. at the First Congregational Church in Oak Park. Despite the time crunch, Ginny was able to put together a beautiful wedding party.

Robert and Virginia's Wedding Party (Courtesy Virginia S. Davis)

Robert and Virginia, 28 November 1942
First Congregational Church, Oak Park, Illinois

(Courtesy Virginia S. Davis)

Virginia next to the car they drove to Florida.(Photo Courtesy Virginia S. Dvais.)

After the wedding the newlyweds drove to Orlando where Robert was stationed and began their new life together in a small house at 1209 E. Kaley. While Robert worked, Ginny set up the home they would share. The couple was young and in love with their whole lives ahead of them. Or so they thought.

Coming Home Notes

[16] "Cicero to Honor Own Flying Tiger: Day of Tribute to all Fighters is Next Sunday." *Chicago Daily Tribune (Chicago)*, 26 July 1942, p. W1; digital images, ProQuest (http://www.proquest.umi.com : accessed 3 June 2010), Historical Newspaper Collection.

[22] "Bob Brouk Air Crash Victim," *(Berwyn) Berwyn Life*, 20 December 1942, p. 1, col. 5.

[23] "Committee Works on Plans for Brouk Day Celebration," *(Berwyn) Berwyn Life*, 22 July 1942, p. 1, col. 2.

[24] "Sees Permanent U.S. Air Force of Two Million Men," *Chicago Daily Tribune (Chicago)*, 23 July 1942, p. 6; digital images, ProQuest (http://www.proquest.umi.com : accessed 13 October 2010), Historical Newspaper Collection.

[25] "Complete Radio Programs and Highlights for Today," *Chicago Daily Tribune (Chicago)*, 22 July 1942, p. 12; digital images, ProQuest (http://www.proquest.umi.com : accessed 13 October 2010), Historical Newspaper Collection.

[26] "Further Fete Plans Honoring Flying Tiger," *(Berwyn) Berwyn Life*, 24 July 1942, p. 1, col. 2.

[27] "Chinese to Honor 'Flying Tigers'," *Chicago Herald American*, 26 July 1942, p. 16, col. 3.

[28] "Complete Radio Programs and Highlights for Today," *Chicago Daily Tribune (Chicago)*, 27 July 1942, p. 10; digital images, ProQuest (http://www.proquest.umi.com : accessed 13 October 2010), Historical Newspaper Collection.

[29] "Complete Radio Programs and Highlights for Today," *Chicago Daily Tribune (Chicago)*, 30 July 1942, p. 8; digital images, ProQuest (http://www.proquest.umi.com : accessed 13 October 2010), Historical Newspaper Collection.

[30] "Complete Radio Programs and Highlights for Today," *Chicago Daily Tribune (Chicago)*, 6 August 1942, p. 12; digital images, ProQuest (http://www.proquest.umi.com : accessed 13 October 2010), Historical Newspaper Collection.

[31] Hawthorne Works Museum, *Hello Charley 1963*, pamphlet (Cicero, IL. : 1963), inside panel 2.

[32] Virginia S. Davis (Phoenix, Arizona) to "Dear Jen", letter, 1 February 2006; information on Robert Brouk; Robert Brouk

Correspondence File; Robert Brouk Book Research Files; privately held by Jennifer Holik, [address held for private use] Woodridge, Illinois, 2006.

[33] "Tells How 'Tigers' Shot Down Japs," *(Berwyn) Berwyn Life,* 31 July 1942, p.13, col. 2.

[34] "It's Bob Brouk Day Today," *(Berwyn) Berwyn Life,* 2 August 1942, p. 1, col. 5.

[35] "Huge Turnout Predicted for Brouk Fete Sunday," *(Berwyn) Berwyn Life,* 31 July 1942, p. 3, col. 1.

[36] "Flying Tiger Will Invade Lions Den, *(Berwyn) Berwyn Life,* 7 August 1942, p. 5, col. 1.

[37] "Czechs to Renew Their Pledge to Free Way of Life," *Chicago Daily Tribune (Chicago),* 9 August 1942, p. W5; digital images, ProQuest (http://www.proquest.umi.com : accessed 13 October 2010), Historical Newspaper Collection.

[38] "Bob Brouk will Join Moose Lodge," *(Berwyn) Berwyn Life,* 9 August 1942, p. 1, col. 7.

[39] "Bob Brouk will Join Moose Lodge," *(Berwyn) Berwyn Life,* 9 August 1942.

[40] "Flying Tiger Gives Vivid Account of His China Air Fights," *The Microphone* 18 (September 1942) : 5.

[41] "Radio Stars Join in Show to Aid Air Recruiting, *Chicago Daily Tribune (Chicago),* 9 October 1942, p. 22; digital images, ProQuest (http://www.proquest.umi.com : accessed 13 October 2010), Historical Newspaper Collection.

[42] "Capt. Bob Brouk," *(Berwyn) Berwyn Life,* 14 October 1942, p. 10, col. 7.

[43] War Department Report of Death 29 December 1942; *Individual Deceased Personnel File*; Military Textual Reference Branch, National Archives, College Park, MD.

[44] Compiled Army Air Force Accident Report, Robert R. Brouk, Captain, 50th Fighter Group, 10th Fighter Squadron, Records of the Army Air Force, p. 20.

[45] Jo Neal, President AVG-FTA, [(E-address for private use),] to Jennifer Holik, email, 13October 2010, "Robert Brouk," Robert Brouk Correspondence File, Robert Brouk Book Research Files; privately held by Jennifer Holik [(E-address) & street address for private use,] Woodridge, Illinois.

[46] Virginia S. Davis to "Dear Jen," 1 February 2006.

[47] Virginia S. Davis (Brouk), "Memoir 1918 – 2010" (MS, Phoenix, Arizona, 2010), p. 139; privately held by Virginia S. Davis, [Address for private use,] Phoenix, Arizona, 2010.

[48] "Flying Tiger's bride-to-be gets license," *(Chicago) The Daily Times,* Chicago, 25 December 1942, p. 20, col. 3.

The Final Flight

Life can turn on a dime and you never know what will happen next. On December 19, 1942, Ginny woke up excited because Robert was due home that morning. He had been training pilots in Tallahassee, Florida for three days. After breakfast, she drove to the airfield to await Robert's return.

Ginny recalled, "I left the house about 9:30 a.m. to pick up Bob who had been on maneuvers the past three days. He was a flight instructor and he and the student pilots were returning from southern Florida in close flying formation. I parked by the base hangars and sat on the hood of our car awaiting his arrival."

Upon arrival at the Orlando Air Base, Robert was the lead plane in a formation of six ships. What a magnificent sight! The planes rose and dove in formation as they practiced maneuvers. Eventually they would be sent overseas to whip the enemy.

The ships in Robert's formation, along with other formations, were simulating a strafing of the Orlando airfield. As the strafing simulation was being attempted, First Lieutenant Sidney O. Kane's plane, flying in the number two position, flew too close to Robert's plane as Robert began to pull up and turn left.

What happened next was inconceivable. Flying just a few hundred feet off the ground, Lt. Kane's wing tip touched Robert's right wing tip. This small action caused Robert's plane to immediately flip upside down, crash, and explode. What remained of his plane was almost unrecognizable as aircraft.

Ginny watched the entire event unfold from the hood of their car.

Witnesses reported Lt. Kane's plane flew on a short distance, gained some altitude and then turned and crashed into a field. He did not survive.

"I watched the planes pass overhead, then turn and prepare for a landing when all of a sudden two planes collided in mid-air and fell to the ground. Shocked, I sat there in silence and just stared until some people came and informed me that one of the planes was Bob's." Ginny recalled.[49]

Ginny was in such shock, her memories of the events after the crash were hazy. It was too much for a 20 year old new bride to process. Moments before the crash she was excitedly anticipating her new husband's return and a day together. But that was not to be. Her blood pressure soared and she was put in bed and packed in ice to bring the pressure down.

The next thing Ginny remembered was that she phoned her in-laws to tell them Robert had died. Still in shock, she did as instructed by the military captain assigned to escort her home, packed a bag and departed for the train station. The rest was a blur.

After the Army Air Corps handled the packing of Robert and Ginny's belongings and shipped them to Chicago. The Chinese robe Robert gave her was not among the possessions returned. She has no idea what happened to it after his death.

The U.S. Army Air Force investigated the accident, Accident Number 43-12-19-12, dated December 19, 1942. The following witness descriptions and crash photos are taken from this Accident Report.[50]

Several Description of Accident and Witness Statements are included in the Accident Report 43-12-19-12. The first account was given by William J. Cummings, Jr. Maj. AC Aircraft Accident Officer, dated 24 December 1942.

On December 19, 1942 at approximately 1220 a P-40E Type airplane A.F. No. 41-24928, piloted by Lt. S.O. Kane, crashed and burned in the vicinity of Kissimmee, Fla after a mid-air collision with another aircraft. The crash proved fatal to the pilots of both airplanes.

> It is found that Lt. Kane was flying in the number two position of a six ship flight which was acting as "top cover" for two other flights that were making a simulated "ground strafing" attack on the Kissimmee, Fla. Airfield. After the attack had been made by other flights the leader of the "top cover" flight gave orders by radio for his flight to execute a "number one" attack on the airfield. This type of an attack is executed with individual airplanes flying in

"trail" with approximately 100-125 yds. between airplanes.

It is further found that during the attack, the airplane flown by Lt. Kane was entirely too close to the airplane leading and that the airplane flown by Lt. Kane struck the leaders airplane during a pull up after the attack. The leader's airplane immediately went out of control and crashed and burned. The airplane flown by Lt. Kane continued in a climb for a short distance, lost the propeller and then crashed and burned after what appeared to be an attempt by the pilot to make a forced landing.

It is the opinion of the Aircraft Accident Board that, in spite of previous orders and instructions, the pilot, Lt. Kane, was not flying in the proper position for the type of formation ordered and was directly responsible for the occurrence of the accident.[51] (sic)

The second account was given by William Couldery, dated 19 December 1942.

I was watching two planes as they dived and came across Kissimmee Airport at an altitude of approximately 45 ft. The two planes were flying one behind the other at an altitude of 45 ft, when the rear plane tried to get ahead of the other one and they collided wing tips. The plane that had been in the rear turned over and dived to the ground within 30 yards from where I was standing. The plane turned wing over wing for approximately 50 ft. before the plane caught on fire, bursting in half, part of the ship continued on for approximately 50 ft., burning all the while. The other plane started gaining altitude and was up to about 200 ft. before it dropped and hit the ground about a mile and a half from where I was standing.

This happened at approximately 11:45 A.M. December 19, 1942, one quarter of a mile south of Kissimmee Airport.[52] *(sic)*

Another account was given by Joseph P. Lemons, 1st Lt, Air Corps, 10th Fighter Squadron.

On December 19, 1942, about 12:10 o'clock, Captain Brouk with a formation of six ships, of which I was flying number four, made a strafing attack on Kissimmee Air Port. The formation approached with Lt Kane flying very close to Captain Brouk. I was about three hundred yards behind Lt Kane directly in trail. As the two planes ahead pulled up to clear the hangar one of them suddenly turned on its back and crashed; then the other climbed up to about six hundred feet and the propeller fell off. The second ship then apparently assumed a normal glide toward a field almost directly ahead. When the plane reached an altitude of about two hundred feet the pilot made an extremely steep turn, toward a better field, pulled his nose up in the turn and fell off on the left wing and plunged straight into the ground. Both planes burst into flames as they crashed. I did not see the two planes collide in mid-air.[53] *(sic)*

A fourth account was given by Walter J. Koraleski, Jr., 1st Lieut, Air Corps, 10th Fighter Squadron stating,

On December 19, 1942, 17 P-40E-1's of the 10th Fighter Squadron took off at approximately 11:15 from Dale Mabry Field, Tallahassee for the Orlando Air Base. There were three flights of six, six and five ships. Captain Brouk in airplane 956 was leading our flight of six with Lieut. Kane in plane 928, his wing man and myself in number 3 position. We were air support for the other two flights which were to attack Kissimmee.

When we came to Kissimmee the two flights made their attack and left. Captain Brouk then gave us the order for No. 1 attack and we peeled off from 4000 feet. The first attack was from Southwest to northeast. Captain Brouk then pulled-up and turned for another attack at the hangar from North to South. On making this last turn I noticed Lieut. Kane was quite close to Captain Brouk, approximately one to two ship lengths. After passing the hangar Captain Brouk pulled up and started to turn to the left. Lieut. Kane, who was on Captain Brouk's right wing turned also, and suddenly his propeller hit the underside of Captain Brouk's ship. At this time they were at an altitude of approximately 100 feet and I was approximately 200 yards behind them. Pieces flew off at the impact and then the two ships separated. Captain Brouk's ship, which was still in its bank, did a sort of one half slow roll as it went down and landed almost upside down. The ship burned on impact with the ground and the flames shot along the ground for what looked to be hundreds of feet.

Lieut. Kane's ship didn't seem to be damaged too much. The prop was stopped and all three blades were bent back. By this time I was approximately 100 yards behind him and could see that his canopy was open and that meanwhile he had pulled up to about 500 feet. Suddenly the prop fell off. Lieut. Kane continued to glide straight ahead and seemed to have control of the plane. He continued south until passing the cement highway and now was heading toward a field that had either small trees or brush in it. To his left was a clear green field and when at about 200 feet he turned toward that field. He completed about 100 degrees of that turn when suddenly the ship dived for the ground and rolled almost on its back. It hit the ground and flames shot up immediately.

I called Captain Kiser on the radio and told him what had happened. He returned to Kissimmee and I led the remaining ships of our flight to Orlando and landed.[54] *(sic)*

Army Air Force Accident Report Microfilm - Lt. Kane's Crash Photograph 1[55]

Army Air Force Accident Report Microfilm - Lt. Kane's Crash Photograph 2[56]

Army Air Force Accident Report Microfilm - Robert's Crash Photograph 1[57]

Arm
y Air Force Accident Report Microfilm - Robert's Crash Photograph 2[58]

Army Air Force Accident Report Microfilm - Robert's Crash Photograph 3[59]

Army Air Force Accident Report Microfilm - Robert's Crash Photograph 4[60]

The Final Flight Notes

[49] Virginia S. Davis (Scharer), "Memoir 1918 – 2010," 145.

[50] Compiled Army Air Force Accident Report, Robert R. Brouk, p. 19.

[51] Compiled Army Air Force Accident Report, Robert R. Brouk, page 4.

[52] Ibid, page 10.

[53] Ibid, page 12.

[54] Ibid, page 24.

[55] Ibid, page 15.

[56] Ibid, page 15.

[57] Ibid, page 31.

[58] Ibid, page 31.

[59] Ibid, page 32.

[60] Ibid, page 32.

The Family's Grief

Following the crash, Virginia quiet and withdrawn, accompanied Robert's remains, which were sent by rail from Florida to Chicago.

Robert's body laid in state for two and a half days from December 22 – 24, 1942 at the Chrastka Funeral Home in Cicero, Illinois.

Remembering the kind, quiet, hero who had recently married Western Electric's sweetheart, thousands of people swarmed the funeral home and stood in line for hours to pay their respects to Cicero's Hometown Hero.

Ginny recalled standing for hours in the funeral home as people passed through. Quiet, strong, and detached, she accepted their condolences. It was improper to grieve or break down in front of all those people because she was on display for all to see.

Because of her fame as a "Hello Charley" girl and being the widow of a famous Flying Tiger, Ginny was no longer Ginny Brouk. She would forever be remembered as *The Tiger's Widow*. Her fame, as a result, would be used by the newspapers and Army for years to come as part of the patriotic fervor.

In keeping with military tradition, a full military funeral service was held with Honor and Color Guards provided by soldiers from Fort Sheridan and local Veterans' organizations. These were many of the same organizations that participated in "Bob Brouk Day" just a few months earlier.[61]

During the funeral service, "Dr. Albert Buckner Coe, pastor of the First Congregational Church in Oak Park, delivered a brief eulogy in English after Otto Pergler had spoken for a short time in the Czech tongue."[62] Dr. Coe, who married Robert and Virginia just weeks before, spoke about the strength of Robert's character in life and death; how Robert loved freedom and did his part to fight for it; and how strong and courageous Robert was stating, "Captain Brouk was fearless in life, and we're sure, fearless in death."[63]

The Berwyn Life reported on the services and funeral. The paper described the emotions of the young servicemen's wives and their parents, who attended the services, as deeply grieved; the emotions of the fathers as sober, yet understood the anger and sadness that Robert's father must have been feeling; and mothers whose hearts went out to Robert's mother and a whole community was in mourning for a heroic man lost so young.

These same parents who attended the wake and funeral were no doubt thinking about their boys fighting in Europe or the Pacific. Would their sons come home unharmed, unchanged? Or would they sleep forever in foreign soil among their brothers in arms? These thoughts cast more darkness on an already dark day.

Upon conclusion of the ceremony at the funeral home, Robert's casket was transported to Woodlawn Cemetery in Forest Park, Illinois. *The Berwyn Life* reported it took three automobiles to transport the many floral arrangements sent to the funeral home for Robert; there were more than a dozen American and Veteran Flags as part of the funeral service; and six Army Air Corps officers served as pall bearers.[64]

Ginny and Robert's parents sat quietly near the grave site where the casket was placed and listened as Fort Sheridan post chaplain, Erling C. Grevstad officiated the graveside service. Following the eulogy, a wave of sadness descended upon the crowd as a soldier mournfully played Taps. Then the boom of the guns, as a salute was fired, woke them from their quiet reverie as they prepared to depart.

Robert was laid to rest in the Birchwood Section of the cemetery while his family silently grieved for the man who was a modest, heroic, honorable husband, son and brother.

Ginny described the whole funeral process as, "the most devastating of anything I had ever had to go thru. I passed out while watching them lower the casket into the ground. I was very young and completely unprepared for such a trauma. After all, I was still on my honeymoon."[65]

The following Tuesday, *The Berwyn Life* reported the Berwyn city council adopted a resolution to be sent to the wife and parents of Robert. It said in part, "Our country is richer because of his bravery, and a world peace is nearer because of his labors."[66]

Robert's Grave. (Author's photo.)

The Family's Grief Notes

[61] "Brouk Rites are Thursday," *(Berwyn) Berwyn Life,* 22 December 1942, p. 1, col. 6.

[62] "Thousands see Brouk Rites," *(Berwyn) Berwyn Life,* 27 December 1942, p. 1, col. 6.

[63] "Thousands see Brouk Rites," p. 1, col. 6.

[64] "Paying Final Tribute to Captain Brouk," *(Berwyn) Berwyn Life,* 29 December 1942, p. 1.

[65] Virginia S. Davis (Scharer), "Memoir 1918 – 2010," 146-147.

[66] "Thousands see Brouk Rites," *(Berwyn) Berwyn Life,* 27 December 1942, p. 1, col. 6.

Robert Is Remembered

Robert Brouk. (Photo courtesy Virginia S. Davis.)

Robert's short life affected many people, and he was honored in numerous ways. After Robert's death, Virginia received a letter from the War Department sending their condolences on his tragic accident.

WAR DEPARTMENT
COMMANDING GENERAL, ARMY AIR FORCES
WASHINGTON

January 4, 1943

My dear Mrs. Brouk:

It is with deep regret that I have learned of the untimely death of your husband, Captain Robert Ralph Brouk, on December 19, 1942, in an airplane accident at Kissimmee Airport, Kissimmee, Florida.

It has come to my attention that Captain Brouk was not only a highly skilled pilot, but an officer of rare courage and great charm who captured the affectionate respect of all his associates.

Captain Brouk early proved his valor and fine philosophy of life by serving with the magnificent "Flying Tigers" in the Far East before his transfer to the Army Air Forces. Later, as an instructor at the Orlando Air Base in Florida, he performed important work with great efficiency and was in every respect a credit to his command.

I hope the memory of the gallantry and heroism displayed by your husband in serving his Country, and other nations fighting for the same cause, will be a measure of comfort in your grief, and, with the passage of time, will alleviate your great sorrow.

My deepest sympathy to you and to other members of the family.

Very sincerely,

H. H. ARNOLD,
Lieutenant General, U.S. Army,
Commanding General, Army Air Forces.

Mrs. Robert R. Brouk,
1209 E. Kaley Avenue,
Orlando, Florida.

In early March of 1943, Madame Chiang Kai-Shek visited Chicago. Many concerts and events were held in her honor during the weeks she visited the city. Virginia received an invitation to attend a reception and tea on 12 March at the Palmer House in Chicago.

Chang Lok Chen

Consul General for China

requests the pleasure of the company of

Mrs. Virginia S. Brouk

at a reception and tea in honor of

Madame Chiang Kai-Shek

at the Palmer House

on Friday afternoon, March 12th - 1943

from five to seven

R. S. V. P.
To Chang Lok Chen
201 North Wells Street
Chicago

Invitation to tea (Courtesy Virginia S. Davis)

Uniforms to ensure that if they were shot down, the Chinese people would assist the pilots.[68]

Virginia remembered Madame Chiang Kai-Shek as a very cultured and charming woman who remembered Robert and his 3rd Squadron of the Flying Tigers.[67]

The Hawthorne Works Newsletter, *The Microphone*, featured Virginia in an article about her meeting with Madame Chiang Kai-Shek and explained that the large scarf she is wearing was in appreciation of Robert's service in China. The small scarf-like item on the right side of the photo is a cloth that was sewn on the back of all of the Flying Tiger's

Virginia wearing silk scarf
(Courtesy Virginia S. Davis)

In December 8, 1996 the Flying Tigers were awarded the Distinguished Flying Cross on, for extraordinary achievement while participating in combat as part of the American Volunteer Group in China.

I contacted the Flying Tiger Association and asked who would receive Robert's medal and citation. I explained I was one of the only family members left, as his parents and brother were deceased. I was fortunate and received Robert's award and its accompanying citation.

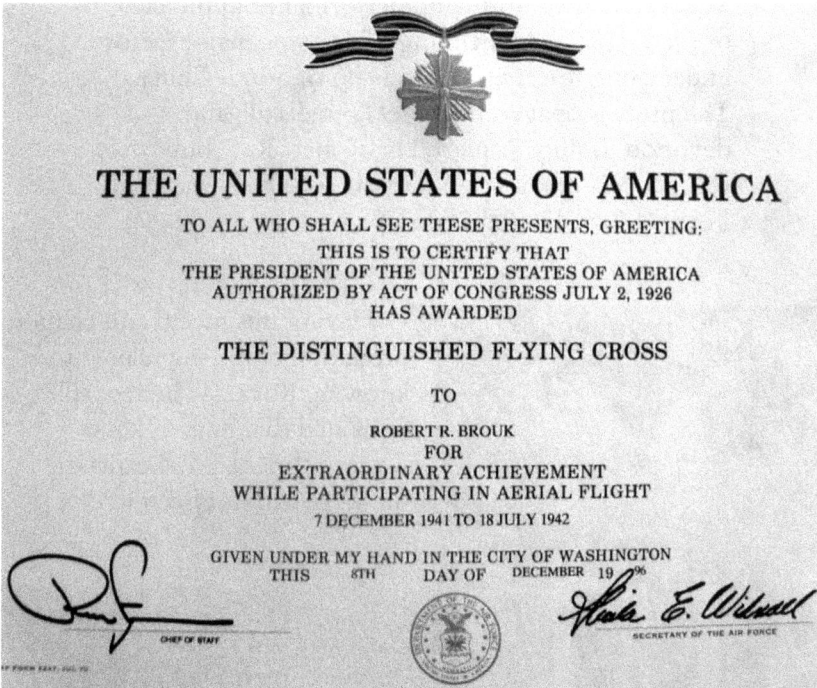

THE UNITED STATES OF AMERICA

TO ALL WHO SHALL SEE THESE PRESENTS, GREETING:

THIS IS TO CERTIFY THAT
THE PRESIDENT OF THE UNITED STATES OF AMERICA
AUTHORIZED BY ACT OF CONGRESS JULY 2, 1926
HAS AWARDED

THE DISTINGUISHED FLYING CROSS

TO

ROBERT R. BROUK
FOR
EXTRAORDINARY ACHIEVEMENT
WHILE PARTICIPATING IN AERIAL FLIGHT

7 DECEMBER 1941 TO 18 JULY 1942

GIVEN UNDER MY HAND IN THE CITY OF WASHINGTON
THIS   8TH   DAY OF   DECEMBER 19 96

CHIEF OF STAFF

SECRETARY OF THE AIR FORCE

The Citation awarded stated,

> The American Volunteer Group, the Flying Tigers, compiled an unparalleled combat record under extremely hazardous conditions. This volunteer unit conducted aggressive counter-air, air-defense, and close air support operations against a numerically superior enemy force occasionally 20 times larger. Members of the All Volunteer Group destroyed some 650 enemy aircraft while suffering minimal losses. Their extraordinary performance in the face of seemingly overwhelming odds was a major factor in defeating the enemy's invasion of South China. The professional competence, aerial skill, and devotion to duty displayed by Robert R. Brouk reflect great credit upon himself and the Armed Forces of the United States.[69]

Having this medal and citation brought me one step closer to knowing Robert. The fact still remained that no one knew where happened to Ginny and his story had not been told.

Charles Bond, Jr., Vice Squadron Leader in the AVG, remembered Robert in a letter dated 20 June 1997.

> The only time I got to be with him for any length of time was when we both had been wounded in combat and were in the hospital together in May 1942, in Kunming, China. He had been strafed while on the ground – hit in the legs with several bullets by a Japanese attack. I had been shot down and pretty badly burned. We swapped stories for several days. A very likeable guy. When we were released from the hospital we reported back to our respective squadrons – he to the 3rd and I to the 1st Sq.[70]

Charles Baisden, an Armorer in the AVG, remembered Robert as a man with a great sense of humor with whom he could have friendly conversations about air gunnery.

> We knew each other being in the same squadrons in the states (Langley Field, VA, and Mitchel Field, NY) in 1941 before the AVG Our time at Groton [Air Gunnery Training in Groton, CT] was a matter of weeks but do remember it was here we first learned about Chennault and defending Burma Road. Of interest I saw Bob make a belly landing in an old Martin B10 bomber at Mitchel when they could not get one of the landing gear down.[71]

Charles elaborated on the B10 Bomber landing in a post on the American Volunteer Group's message board in 2003 saying,

> When I was checking out of Mitchel Field about to leave for the AVG I saw an old Martin B10 Bomber which we used for towing targets circling the field with one [landing] gear down and one up. It was Bob Brouk who had been detailed to fly some of our crew chiefs around the area in order to get air time for them to draw some flight pay. As we watched, several people bailed out except for one guy who got his foot caught in the bombardier's hatch and he made a tour of the field hanging upside down. Finally worked his foot free and

parachuted safely to the ground. Turned out it was Charley Kenner who was later one of the 3rd Squadron crew chiefs in the AVG. After the passengers had jumped Bob landed the old crate safely but rather messed up.[72]

Charles also wrote about an incident in early 1942 where Robert was involved in an air attack with the Japanese.

In the early spring of 1942 we bugged out of Loiwing and went to Mengshi just northeast of us. When our planes came in later I started rearming our aircraft. Bob Brouk told me he had come down through an overcast [sky] and found himself alone as a tail end Charlie, in a flight of Japanese fighters. He said they never saw him and he edged up to one of them and fired a very short burst [at] point blank range. The Jap rolled over and went down. Bob said he got the hell out of there. He was also very surprised when I gave him a fired bullet that had gone into his right wing and lodged in one of his .30 caliber wing gun ammunition containers.[73]

It seems someone wanted Robert's story told, for Robert's widow, Ginny, emailed me at the end of 2005. Apparently her grandson saw one of my posts about Robert on a Flying Tigers message board. We corresponded over several months, and Ginny shared some photographs, their love story, and Robert's war diary with me. She told me about her role in Robert's life and about what happened to her after his death.

Ginny and I lost touch for a while until early 2010 when I decided it was time to write Robert's story. I had enough information, his war diary, and with a little additional research, could tell his story. And the best part, I now knew what happened to Ginny after Robert's death.

Would you like to hear her remarkable story of life after death? Please look for the book *The Tiger's Widow*, available early 2014.

Robert is Remembered Notes

[67] Virginia S. Davis (Scharer), "Memoir 1918 – 2010," 154.

[68] Virginia S. Davis (Scharer), "Memoir 1918 – 2010," 155.

[69] Robert R. Brouk Distinguished Flying Cross Medal and Citation; privately held by Jennifer Holik-Urban, [Address for private use] Woodridge, Illinois.

[70] Charles R. Bond, Jr. (McKinney, Texas) to "Dear Ms. Jennifer Tharp", letter, 20 June 1997; information on Robert Brouk; Robert Brouk Correspondence File; Robert Brouk Book Research Files; privately held by Jennifer Holik, [address held for private use] Woodridge, Illinois, 1997.

[71] Charles Baisden, Savannah, Georgia [(E-address for private use),] to Jennifer Holik, email, 7 May 2010, "Robert Brouk," Robert Brouk Correspondence File, Robert Brouk Book Research Files; privately held by Jennifer Holik [(E-address) & street address for private use,] Woodridge, Illinois.

[72] Charles Baisden, "Bob Brouk," *Flying Tigers: American Volunteer Group*, discussion list, 18 December 2003 (www.flyingtigersavg.com/ : accessed 8 May 2010).

[73] Baisden, "Bob Brouk," *Flying Tigers: American Volunteer Group*, 18 December 2003.

[84] Robert R. Brouk Distinguished Flying Cross Medal and Citation; privately held by Jennifer Holik, [Address for private use] Woodridge, Illinois.

Bibliography

"1910 United States Federal Census, Roll T624_254 : 1910." Database *Ancestry.com*. http://www.ancestry.com : 2010.

"1920 United States Federal Census, Roll T625_388 : 1920." Database *Ancestry.com*. http://www.ancestry.com : 2010.

"1930 United States Federal Census, Roll 498 ; 1930." Database *Ancestry.com*. http://www.ancestry.com : 2010.

Baisden, Charles. *Flying Tiger to Air Commando*. Atglen, PA: Schiffer Military History, 1999.

Bond, Charles R., Jr. and Terry Anderson. *A Flying Tiger's Diary*. Texas: Texas A&M University Press, 1984.

Brouk, Robert. Photograph. 1935. 1935 Morton Yearbooks. Morton East High School Archives. Cicero, Illinois.

Chennault, Anna. *Chennault and the Flying Tigers*. New York: Paul S. Eriksson, Inc., 1963.

Davis, Virginia S. (Scharer). "Red Hat Mommas of Pinnacle Peak." MS. Phoenix, Arizona, 2010. Privately held by Virginia S. Davis, [Address for private use,] Phoenix, Arizona. 2010.

Davis, Virginia S. (Scharer). "Memoir 1918 – 2010." MS. Phoenix, Arizona, 2010. Privately held by Virginia S. Davis, [Address for private use,] Phoenix, Arizona. 2010.

Davis, Virginia S. (Scharer). Photographs. ca. 1941 - 1942. Privately held by Virginia S. Davis, [Address for private use,] Phoenix, Arizona. 2010.

Florida Office of Vital Statistics. Death Certificates. Bureau of Vital Records, Tallahassee.

Flying Tigers: American Volunteer Group, discussion list, 2003-2010. http://www.flyingtigersavg.com/ : 2010.

Hawthorne Works Museum. *Hello Charley 1963*. Pamphlet. Cicero, IL.: 1963.

Hawthorne Works Museum. *Microphone*, September 1942. Newsletter. Cicero, IL : 1942.

Hawthorne Works Museum. *Microphone*, December 1942. Newsletter. Cicero, IL : 1942.

Holik, Jennifer, transcriber. "What's Next Diary of Robert R. Brouk From April 1941 – July 4, 1942." MS. 1941-1942. Copy held by Jennifer Holik, [Address for private use,] Woodridge, Illinois. 2010.

Hotz, Robert, Editor. *Way of a Fighter, The Memoirs of Claire Lee Chennault.* NY: G.P. Putnam's Sons, 1949.

Illinois, Berwyn. *The Berwyn Life*, 1938–1945. Scattered issues.

Illinois, Chicago, City of. Probate case files. Circuit Clerk's Office, Chicago.

Illinois, Chicago. *Chicago Daily Tribune*, 1930–1943.

Illinois, Chicago. *The Herald-American*, 1942. Scattered issues.

Illinois, Cicero. *The Cicero Life*. 1943. Scattered issues.

Illinois, Cicero. *Morton Collegian*, 1936-1937. Scattered issues.

"J. Sterling Morton Year Book 1935." Database *Ancestry.com*. http://www.ancestry.com : 2010.

Losonsky, Frank and Terry Losonsky. *Flying Tiger A Crew Chief's Story*. Atglen, PA: Schiffer Military History, 1996.

Military, Compiled Army Air Force Accident Report 43-12-19-12. World War II. Records of the Army Air Force, 1942.

Mireless, Anthony J. *Volume 1: Introduction, January 1941 - June 1943 "Fatal Army Air Forces Aviation Accidents in the United States, 1941-1945".* Jefferson, NC: McFarland & Company, Inc., 2006.

Morton High School Archives. Yearbook. ca. 1935. Held by Morton High School, [2400 Home Avenue,] Berwyn, Illinois, 1935.

Morton Junior College Archives. Pioneer Yearbook. ca. 1936. Held by Morton Junior College Library, [3801 S. Central Avenue,] Cicero, Illinois, 1936.

Morton Junior College Archives. Pioneer Yearbook. ca. 1937. Held by Morton Junior College Library, [3801 S. Central Avenue,] Cicero, Illinois, 1937.

Moser, Don, and the Editors of Time-Life Books. *China-Burma-India*. Time-Life Books, Inc., 1978.

"New York Passenger Lists, Roll T715_99: 1820-1957." Database *Ancestry.com*. http://www.ancestry.com : 2010.

"Old Ship Picture Galleries." Website http://www.photoship.co.uk : 2010

Pistole, Larry M. *The Pictorial History of the Flying Tigers*. VA: Publisher's Press, Inc., 1981.

Red Hat Mommas of Pinnacle Peak, 2005.

Robert Brouk Artifact Collection. Owned by Jennifer Holik, [Address for private use] Woodridge, Illinois. 2010.

Robert Brouk Book Research Files. Privately held by Jennifer Holik, [Address for private use] Woodridge, Illinois. 2010.

San Diego Air and Space Museum Library and Archives. Photographs. 1942. Held by the San Diego Air and Space Museum Library and Archives, [2001 Pan American Plaza,] San Diego, California. 2010.

Schultz, Duane. *The Maverick War*. New York: St. Martin's Press, 1987.

Shilling, Erik. *Destiny: A Flying Tiger's Rendezvous with Fate*. Privately Published, 1993.

Smith, R.T. Photograph. 1942. Digital image. Privately held by Brad Smith, [Address for private use,] Berkeley, California. 2010.

"Social Security Death Index." Database *Ancestry.com*. http://www.ancestry.com : 2010.

United States Army. *Individual Deceased Personnel File*. Military Textual Reference Branch. National Archives, College Park, MD.

Index

About the Author

Jennifer Holik is a genealogical research professional and the owner of Generations. She has a BA in History from Missouri University of Science and Technology.

Jennifer has over eighteen years of research and writing experience. Her most current endeavor is building a genealogy department and programming at Casa Italia – The Italian Cultural Center, in Stone Park, IL, a Chicago suburb. In the fall of 2013 Casa Italia will begin monthly genealogy programming and writing groups. Additional programming will develop in 2014.

Her *Branching Out* kids' genealogy curriculum books have been used by genealogy societies, libraries, homeschool settings and public schools. She writes several blogs and has authored articles for local and national genealogical publications. She writes for the website Archives.com and the genealogy digi-mag, The In-Depth Genealogist.

Jennifer lectures in the Chicagoland area on ways to use technology with genealogy, tracing the history of your home, teaching genealogy to kids and teachers, and finishing the stories of your military ancestors.

Learn more about Generations by visiting http://generationsbiz.com.

Contact the author at jenniferholik@generationsbiz.com

Books Available from Generations

You can view all my latest products on my website:
http://generationsbiz.com/products.html

Stories of the Lost
Due out Winter 2014

The Tiger's Widow
Due out Winter 2014

Branching Out: Genealogy for 1st – 3rd Grade Students Lessons 1-30
https://www.createspace.com/3854119

Branching Out: Genealogy for 4th – 8th Grade Students Lessons 1-30
https://www.createspace.com/3854226

Branching Out: Genealogy for High School Students Lessons 1-30
https://www.createspace.com/3854252

Branching Out: Genealogy Lessons for Adults
https://www.createspace.com/3852836

Engaging the Next Generation: A Guide for Genealogy Societies and Libraries
https://www.createspace.com/3802454

Legacy QuickGuide: World War I and World War II Military Records (PDF)
https://www.legacyfamilytreestore.com/ProductDetails.asp?ProductCode=QDWW&Click=114310

Legacy QuickGuide: Italian Genealogy (PDF)
https://www.legacyfamilytreestore.com/ProductDetails.asp?ProductCode=QDITALIAN&Click=114310

Legacy QuickGuide: Land and Property Records
Card –
https://www.legacyfamilytreestore.com/ProductDetails.asp?ProductCode=Q_LANDPROP&Click=114310

PDF -
https://www.legacyfamilytreestore.com/ProductDetails.asp?ProductCode=QDLANDPROP&Click=114310

Legacy QuickGuide: Genealogy for Kids
Card -
https://www.legacyfamilytreestore.com/ProductDetails.asp?ProductCode=Q_CHILDREN&Click=114310

PDF -
https://www.legacyfamilytreestore.com/ProductDetails.asp?ProductCode=QDCHILDREN&Click=114310

Legacy QuickGuide: Missouri Genealogy Records (PDF)
https://www.legacyfamilytreestore.com/ProductDetails.asp?ProductCode=QDMISSOURI&Click=114310

Legacy QuickGuide: Indiana Genealogy Records (PDF)
https://www.legacyfamilytreestore.com/ProductDetails.asp?ProductCode=QDINDIANA&Click=114310

Legacy QuickGuide: Arkansas Genealogy Records (PDF)
https://www.legacyfamilytreestore.com/ProductDetails.asp?ProductCode=QDARKANSAS&Click=114310

Legacy QuickGuide: Iowa Genealogy Records (PDF)
https://www.legacyfamilytreestore.com/ProductDetails.asp?ProductCode=QDIOWA&Click=114310

Legacy QuickGuide: Michigan Genealogy Records (PDF)
https://www.legacyfamilytreestore.com/ProductDetails.asp?ProductCode=QDMICHIGAN&Click=114310

To Soar with the Tigers
Paperback available through CreateSpace
https://www.createspace.com/3549200
EPub format available on Barnes and Noble's Nook and Amazon's Kindle

Genealogy Tip Sheets
Tip sheets can be purchased off the author's website:
http://generationsbiz.com/products.html

Using Excel for Genealogical Research
EPub format available on Lulu.com, Barnes and Noble's Nook and Amazon's Kindle

Locating Chicago Property Records
EPub format available on Lulu.com, Barnes and Noble's Nook and Amazon's Kindle

www.ingramcontent.com/pod-product-compliance
Lightning Source LLC
Chambersburg PA
CBHW052211270326
41931CB00011B/2311